AF734706

Science Ltd.

Science Ltd. examines the evolving relationship between scientific knowledge, technology, and society in an era where deep-tech companies are increasingly driven by discoveries rather than market demand. Written from the perspective of metascientific entrepreneurs, this timely work explores how scientific research struggles to balance profit with truth, individual objectives with collective values, and human creativity with machine capabilities. This book addresses the undisputed influence of science in contemporary society while questioning the sustainability of current research paradigms.

Science Ltd. tackles critical operational questions about modern science: How is scientific knowledge produced and prioritized? How are research funds distributed? What constitutes effective dissemination and application of discoveries? Chapters examine the journey from laboratory innovation to commercial implementation while exploring two fundamental challenges facing the scientific community. First, the urgent need to reform scientific practices, particularly by rethinking academic publishing and research funding models to create more open and sustainable science. Second, the complex prospect of human-machine collaboration in research, moving beyond traditional instruments to AI assistants with capabilities that exceed human limitations. Enriched with expert interviews and real-world anecdotes, this book provides both practical insights and speculative analysis of science's automated future.

This accessible work is designed for researchers, entrepreneurs, policymakers, and anyone curious about the intersection of science and technology. It serves academics seeking to understand contemporary research dynamics, industry professionals navigating the deep-tech landscape, and general readers interested in how scientific discovery shapes society. This book offers valuable perspectives for those involved in research funding, science policy, and technology transfer, while remaining engaging for anyone concerned about the future of human-driven versus machine-assisted scientific innovation.

Science Ltd.

Research Enterprise in the Age of Machines

Andrea Borghini and Simone Severini

CRC Press
Taylor & Francis Group
Boca Raton London New York

CRC Press is an imprint of the
Taylor & Francis Group, an **informa** business

Designed cover image: The cover image has been illustrated by Enzo Ragazzini.

First edition published 2026
by CRC Press
2385 NW Executive Center Drive, Suite 320, Boca Raton FL 33431

and by CRC Press
4 Park Square, Milton Park, Abingdon, Oxon, OX14 4RN

CRC Press is an imprint of Taylor & Francis Group, LLC

© 2026 Andrea Borghini and Simone Severini

SCIENZA S.p.A. by Andrea Borghini and Simone Severini

© 2024 by Mondadori Education S.p.A, Milano, Italy

Tutti i diritti riservati – All rights reserved

www.mondadorieducation.it

Reasonable efforts have been made to publish reliable data and information, but the author and publisher cannot assume responsibility for the validity of all materials or the consequences of their use. The authors and publishers have attempted to trace the copyright holders of all material reproduced in this publication and apologize to copyright holders if permission to publish in this form has not been obtained. If any copyright material has not been acknowledged please write and let us know so we may rectify in any future reprint.

Except as permitted under U.S. Copyright Law, no part of this book may be reprinted, reproduced, transmitted, or utilized in any form by any electronic, mechanical, or other means, now known or hereafter invented, including photocopying, microfilming, and recording, or in any information storage or retrieval system, without written permission from the publishers.

For permission to photocopy or use material electronically from this work, access www.copyright.com or contact the Copyright Clearance Center, Inc. (CCC), 222 Rosewood Drive, Danvers, MA 01923, 978-750-8400. For works that are not available on CCC please contact mpkbookspermissions@tandf.co.uk

For Product Safety Concerns and Information please contact our EU representative GPSR@taylorandfrancis.com. Taylor & Francis Verlag GmbH, Kaufingerstraße 24, 80331 München, Germany.

Trademark notice: Product or corporate names may be trademarks or registered trademarks and are used only for identification and explanation without intent to infringe.

ISBN: 9781041148876 (hbk)
ISBN: 9781041131069 (pbk)
ISBN: 9781003676614 (ebk)

DOI: 10.1201/9781003676614

Typeset in Minion
by codeMantra

To Folco, Lelio, and Lia.

Contents

Preface

> The saddest aspect of life right now is that science gathers knowledge faster than society gathers wisdom.
>
> *Isaac Asimov*

> But, it being my intention to write a thing which shall be useful to him who apprehends it, it appears to me more appropriate to follow up the real truth of the matter than the imagination of it; for many have pictured republics and principalities which in fact have never been known or seen, because how one lives is so far distant from how one ought to live, that he who neglects what is done for what ought to be done, sooner effects his ruin than his preservation; for a man who wishes to act entirely up to his professions of virtue soon meets with what destroys him among so much that is evil.
>
> *Niccolò Machiavelli*

> There is nothing more deceptive than an obvious fact.
>
> *Arthur Conan Doyle*

The endeavors to advance scientific knowledge in conjunction with machines have grown exponentially in every respect since 2022—amounts of funding, number of people involved, results in disparate fields of research, media attention etc. In 2024 or 2025, all the principal private companies and public institutions, across the globe, have set up ambitious efforts and plans to speed up these endeavors and outpace competitors. As we release this English edition four years after we first conceived of our book, in early 2022, it is necessary to briefly contextualize it.

The bulk of the manuscript was written between October 2022 and December 2023; at this time, we also held conversations with colleagues that the reader will find in the text. Revisions to the original Italian edition were made during the first half of 2024.

In September 2024, we began work on an English edition. We have updated and revised several passages. But we left the core of the ideas intact. On the one hand, the premise for writing a sequel book may have already been there—holding *today* the conversation we had two or three years ago, between ourselves or with colleagues, would likely lead to evoking additional scenarios. On the other hand, we opined that our approach was still valid, as it moves from first principles to interrogate *foundational and underlying* processes that shape scientific innovation, putting forward strategies and values to best enable it.

About the Authors

Andrea Borghini is Associate Professor in the Department of Philosophy at the University of Milan.

Simone Severini is a Distinguished Engineer at Google and Professor of Physics of Information at UCL.

Acknowledgments

A fundamental contribution to this essay came from all the people who, during the writing phase, dedicated their time to answering our questions and who, with their availability and competence, showed us new perspectives.

We exchanged audio recordings, held video conferences with every possible tool made available to us by technology, made telephone calls (as they used to do), sent e-mail messages, and drank coffee while sitting at a table, something which we hope will never go out of fashion. All with the aim of gathering opinions from people with different experiences, but too often without the possibility of having a chat in person for logistical reasons.

Many of their names are mentioned in the text, but we would like to thank them explicitly: Marco Argenti, Andrea Basso, Polina Bayvel, Ginestra Bianconi, Samuel Braunstein, Francesco Caravelli, Giuseppe Carleo, Vincenzo Cosenza, Byron Cook, Leonardo De Cosmo, Leonardo de Moura, Paolo Dello Vicario, Valeria de Paiva, Brendan Fong, Luca Foschini, Anna Grassellino, Antonio Gulli, Jesse Han, Muki Haklai, Aram Harrow, Thomas Hartke, Ilyas Khan, Sergei Kalinin, Iordanis Kerenidis, Peter Knight, Alex Kontorovich, Stefano Lami, Nanyan (Dorothy) Li, Luca Lorenzini, Adam Marblestone, Fotini Markopoulou, Raffaele Mauro, Francesca Medda, Ashley Rose Mehlenbacher, Maurizio Petta, Diego Piacentini, Aparna Prabhakar, Giuseppe Ragazzini, Joshua Ramette, Andrea Rocchetto, Graeme Smith, Stefano Soatto, Fabio Veronese, and James Whitfield.

Then there are the people who commented on this book, helping to restructure it overall and in detail, as well as improve the exposition: Silvia Bencivelli, Davide Coero Borga, Roberto Gronda, and Leonardo Quattrucci. In addition to them, we also particularly thank: Eva Filoramo, who revised the original book in Italian with exemplary punctuality and acumen in the phase prior to publication; Eleanor Stanford, who copy-edited the English translation and sensibly guided us in revising it; and

Armando Massarenti, to whom we are indebted for supporting the Italian edition of this book.

The text was born from written and oral exchanges between Andrea and Simone via instant messaging applications, stolen from the most singular daily contexts, at the margins of our jobs, often nine hours apart, given the merciless time zones. The last and most intimate thanks go to Meng and Min.

COVER NOTE

Enzo Ragazzini—*Science Ltd.* Book Cover

The cover of this book employs a technique that has accompanied me for most of my life. I invented it in 1965 while working in the darkroom, and I still use it today with Photoshop. It is a form of graphic interference—a way of playing with patterns, grids, or letters. That same year, with this invention, I held my first exhibition of optical art, entirely photographic, at the Einaudi Bookshop in Rome. The cover is, in fact, one of the images from that exhibition.

In 1965, I was largely ignorant of the art world. I didn't even realize that, in the same month, the famous exhibition *The Responsive Eye* was opening in New York—an event that gave a name, Optical Art, to this kind of work. The gallery owner in Rome was also unaware; his interests were more literary, and he was skeptical of so much abstraction. Without knowing it, he had stumbled upon something new.

Explaining in more detail how the cover of this book was made would, however, be tedious for readers. Instead, I prefer to say a few words about my approach to work.

Today, in my 90s, I am still exploring the same line of research I began in 1965, curious to see where it might lead. At times, I feel like the knight in Bergman's *The Seventh Seal*, who, elegant and fearless, plays chess with Death. At other times, however, I feel like Woody Allen, obsessed with his Jewish origins, agonizingly imagining himself on the fateful list of long-dead great Jews—none other than Jesus Christ and Karl Marx—yet not feeling well at all.

Choosing and improvising are central to my work. How? I'd rather explain by using three anecdotes. First: Apollinaire, a close friend of Picasso, once said that some painters carefully copy their ideas, while others, like Picasso, create something in the process that they themselves do not yet know—they are the first to be amazed by what they have done. Second: In an interview with the English critic David Sylvester, Francis

Bacon said that for him the act of painting resembled a battle against disorder and banality. I quote from memory: he was paralyzed by the blank canvas, and sometimes, to begin, he would smear the canvas—just to start that battle whose outcome he could not foresee. Third: In the delightful book *Zen in the Art of Archery*, the Japanese archery master, when a zealous German student—the one who would later write the book—asked him once again what Zen was, replied: "Forget the bow, the arrows, and the target, and wonderful things will happen."

There is a way to share decisions with our mysterious subconscious—almost bypassing the brain, which is sometimes an overly conventional filter, bound by its own patterns and limitations. The challenge is to let the subconscious decide: this mysterious "emergency" sometimes allows us to achieve a kind of unexpected perfection, given that we are imperfect sapiens.

Prologue

This essay is primarily dedicated to the future generation of *metascience entrepreneurs*. This expression, recently employed by Michael Nielsen and Kanjun Qiu (2022), refers to those who strive to improve the social aspects of science in the broadest sense: where it is done, with whom, by what selection and funding methods, and with what means of collaboration, among other considerations. As far as we know, there are no formal studies dedicated to metascience entrepreneurship yet, though there are several courses on specific sub-areas such as science policy, ethics in science, or technology transfer. After reading this, you might think there's a need to inaugurate one—a conviction that, to be honest, grew within us as we wrote this book.

Scientific knowledge wields an undeniable influence over contemporary society. From the geopolitical impact of the Manhattan Project, linked to atomic knowledge, to the evolution of medicine centered on the discovery of DNA, like it or not, in the past century, technology has advanced significantly under the strong impetus of science. This position has only solidified in recent decades, with growth evident across nearly all parameters: from the number of people and institutions working toward scientific advancement to the myriads of annual publications, the vast array of scientific journals, and the sheer volume of funds invested. These aspects reflect growth driven by the idea that scientific knowledge generates dynamics with social benefits, such as improving life expectancy and quality, enhancing security, protecting the planet, or increasing employment and wealth.

DOI: 10.1201/9781003676614-1

Innovation today is, at least in part, due to that intertwining of science and technology known as *deep tech*. The term was coined in 2014 by Indian journalist Swati Chaturvedi and refers to technologies that bring about change based on scientific discoveries rather than new business models. For instance, inventing a new flying vehicle to connect a remote community instead of expanding the existing railway network (an engineering approach) or offering a discount on commuter passes (a managerial approach). Deep tech thus relies on a distinctive category of enterprises driven by the impulse to discover new things rather than by market demand (Gourévitch et al. 2021), often focusing more on building physical products (hardware) than developing processes (software); these enterprises may be partly supported by public funds. Deep tech is one of the main channels for developing contemporary societies, having already had a significant impact in fields such as photonics, materials science, robotics, electronics, quantum technologies, biotechnology, 3D printing, artificial intelligence, aerospace technologies, new agriculture, augmented and virtual reality, and blockchain (Lunden 2020).

However, innovation comes at a cost. Today, scientific research struggles to find the right balance between profit and truth, between individual goals and collective values, between academic rigor and public dissemination, and between human and machine. Opinions about research, as a consequence, quickly tend to consolidate within seemingly opposing schemas—one for all, *pessimists* v. *optimists*.

Reflecting on the state of science and deep tech today is like maneuvering a sailboat with winds and waves constantly changing direction and strength. To get a sense of the difficulty of navigation, one need to consider only two exemplary cases in information technology: the frenzy of announcements triggered in early 2023 by the advent of large language models in machine learning and the controversies generated by the continuous flow of news on quantum computing results (such as the announcement of the so-called quantum supremacy between 2018 and 2019).

The search for balance is also an internal issue within science itself, experienced firsthand daily by those who practice it. Ginestra Bianconi, Professor of Applied Mathematics at Queen Mary University of London, provided a vivid testimony of this during a recent conversation:

> Science is a deeply personal experience, mediated by the external world. On the one hand, there's my vision of research and the way I experience it; on the other hand, there's the scientific community.

> Doing science is a balance between individual motivation and the drive to explore new topics, driven by curiosity, and the reaction of the scientific community to what you write. This balance creates the collective knowledge we call science, a knowledge and practice that today the individual cannot acquire or experience in isolation. Because, ultimately, science is not made by individuals but is a collective phenomenon, even though there have been, and hopefully always will be, people who are clearly 'bearers' of ideas and great innovation.

Today, several signs suggest that the "collective phenomenon" Ginestra speaks of, which underlies scientific production, needs to be rethought. Some pillars that regulate it, such as peer review processes and research funding, show significant and serious difficulties. The new frontiers of artificial intelligence, moreover, suggest perhaps unexpected and unexplored ways of collaboration between machines and humans in the scientific enterprise. How can we try to correct the system to make it more effective?

Combining experiences and perspectives, in this book we sought to clarify our opinions, first and foremost, for and with ourselves during the book's writing. We engaged in discussions for about three years, weaving together perspectives and expertise. We took notes as one would in school notebooks, dividing them into five chapters, providing operational—not speculative—definitions, if you'll allow us the term; definitions born from the point of view of those who, day by day, try to do science or technology, without neglecting the natural problems and issues encountered along the way. Piece by piece, we composed a mosaic that gives us a notebook of doing science in the age of machines.

To bring order to the many small exchanges that took place in writing and orally, we followed a method widely used in contemporary philosophy, clearly, thoroughly, and accessibly described by Roberto Casati in his book *Philosophy as Conceptual Negotiation* (2022): we started with the most elementary ideas that characterize science and technology today, both in content and practice; proceeding gradually and drawing on everyday intuitions, however contested and conflicting, we then added pieces, reviewed and reorganized them, and slowly negotiated a more orderly overall picture.

Addressing the topics and challenges discussed here requires the participation of a broad sector of civil society. For this reason, one of our goals was to write a text that could engage a non-specialist audience. Admittedly, this is not a "light" book. But we were keen to create a possible dialogue even with those who, while not directly involved in scientific production, see its importance and, in some way, must reckon with it: teachers, professionals whose work depends on scientific and technological innovation, investors, communicators, curious individuals with varying levels of experience, people whose choices may impact science (policy makers, for example), and so on.

We therefore begin with broad strokes, dedicating the first two chapters to the foundational concepts of "Science" and "Technology." To readers well-versed in these fields, this initial territory may seem elementary. Please consider these opening sections as the groundwork for the subsequent chapters, where we will delve into the pressing and complex issues that define the modern scientific enterprise.

For the same reasons, we hope that some readers will forgive us if, in drafting our notes, we did not always rely on classic texts and authors that might come to mind during reading, such as Michel Foucault or Bruno Latour. We did not do so to hide a debt to the many who have analyzed science and technology in their most subtle details, discussing the future of human societies, the environment, ecology, Earth, or the Universe, nor to distance ourselves from those who have written about agenda-driven science, such as the collective volume *Politicizing Science* edited by Michael Gough (2003), or acute diagnoses like *Science, Truth, and Democracy* by Philip Kitcher (2001). These studies have obviously influenced us, and when possible, we made our debt explicit. It goes without saying that we didn't invent everything we think, but we didn't want to unnecessarily burden some readers.

How did we manage the risk of addressing a current and complex topic? By seeking the most open and shareable tone we had, we did not shy away from the opportunity to include some anecdotes to liven up a point or testify to what is happening out there. But above all, whenever the going got tough because it required a deeper understanding of the topics, we turned to several experts (such as Ginestra Bianconi, mentioned above): people to whom we owe sincere thanks, not only because their generosity indirectly contributed to the creation of this text but also for sharing pleasant moments that inspired us. We spoke with them, interviewed them, and reported their opinions and ideas as faithfully as we could, as one would do in a report to be offered to a global community of scholars and practitioners. The dialogical approach seemed fitting because it gives voice to those who have formed an opinion in the field, serves as a potential means of verifying opinions, and enriches the analysis by suggesting ideas for alternative routes.

In short: in this book, we attempt to think like two ordinary citizens, old enough to remember what scientific research was like, circa 30 years ago, and young enough to be directly interested in (and perhaps even apprehensive about) its future. In Chapter 1, we ask what characterizes scientific knowledge, while in Chapter 2, we define the scope of technological knowledge. In Chapter 3, we try to rethink how the profession of those who produce scientific knowledge is changing and must change. In Chapter 4, we address technology from the perspective of its interaction with science. Finally, in Chapter 5, we attempt to tie the discussion together in view of its discussion and application.

Two results, more than others, emerged. First, the need to renew how science is done, particularly by rethinking the scientific article paradigm and renewing research funding methods. Second, the urgency of rethinking more deeply the perspective of a science done in collaboration with machines, where the latter are not only seen as laboratory tools or calculators but also as assistants, possessing some technical capabilities that go beyond our own.

In this light, what form and meaning should we impart to scientific education and research today? What values should guide a science co-managed with machines, and what role would that science play in society? Putting together a toolkit for metascience entrepreneurs in the age of machines also means moving into the realm of philosophy, political theory, law, and all the disciplines whose main task is to prescribe how certain things should be, rather than predicting or describing. These are, of course, speculative questions for which it is worth seeking not a formally correct answer but an intelligent formulation, connected to principles and norms suited to managing and organizing scientific knowledge. It is important to note that these principles and norms must also include pragmatic observations, such as management and operational recommendations.

Let's begin.

CHAPTER 1

"Science"

Notes on Science as a Dynamic Activity for Describing and Predicting

The relationship between science and technology defines our modern world, yet we rarely pause to consider how it truly works. To begin a thoughtful exploration of this dynamic, we must first lay some groundwork. Our goal is not to offer a simplified "science popularization" or to build a rigid theory of what science and technology are. Instead, our aim is more direct: to present the essential traits that characterize them today, so we can better understand their shared evolution.

This first chapter begins that journey by focusing on science. We will attempt to sketch out a few of its fundamental ideas in a way that is as accessible as possible. Even when the discussion touches on the technical, we will lean on real-world examples and the voices of those working in the field. This is not meant to be an exhaustive or final word on the matter. It is an opening conversation, and we will proceed, as they say, with caution.

1.1 THE PRACTICAL VALUE OF SCIENCE

One of the most common beliefs is that science plays a hegemonic role in culture and society due to its great practical value. This opinion has been well established among people in science since the last century, supported,

DOI: 10.1201/9781003676614-2

for example, by Nobel laureates Robert Millikan (1924) and Melvin Calvin (1962). How can we express this in simple terms?

1.1.1 Exploration and Prediction

A first, powerful idea is that science is fundamentally about exploration. This is an intellectual exploration, of course, taking place in the vast territory of the unknown. As microbiologist John P. Moore emphasized in a critique of the growing bureaucratization of scientific research (published in *Science* in 2007), it is sometimes useful to think that "science is an exploration of the unexpected." It is perhaps no coincidence that the subjects of scientific study today—from stars and planets to microbes and the brain—were once the domain of philosophical speculation. Science continually seeks to push the boundaries of what is known, sometimes revealing scenarios that are unimaginable, inconvenient, or even unwelcome.

However, this exploratory impulse isn't unique to science. Human societies have always had other ways of probing the edges of experience. The visual arts, music, and theater, alongside practices from the contemplative to the culinary, all serve to both describe our world and speculate on what it could be. Exploration, it turns out, is a deeply human trait.

So, if science is not alone in being characterized by the drive to explore, then where does its uniqueness reside? What separates science from other vital forms of discovery and gives it its particular power?

The reply to these questions rests on a second, crucial idea: science is constitutively fueled by the relentless drive to make *predictions*. The great effort and value we place on science would make little sense without this goal. We see it everywhere: meteorologists forecasting the weather, geologists assigning probabilities to earthquakes, and doctors using prognostic tools to choose the right medical treatment. Even fundamental physics is ultimately a search for theories that can predict the behavior of the universe with a certain degree of accuracy.

This isn't to say every scientific statement is a forecast. Much of science is dedicated to careful description and postulation. But for a truly new idea to enter the scientific arena, it must eventually make a specific, testable prediction. It has to make a claim in the form of a hypothesis: "If my idea is correct, then *this* is what we should observe."

This act of prediction is what gives science its practical power, for two key reasons. First, it allows us to make better decisions, cutting through the noise of randomness and giving us a concrete advantage when choosing a path forward. Second, prediction gives us the potential to *act*. It allows us to interact with nature rather than just be subject to it. By knowing what is likely to happen, we gain the choice to either let events unfold or to intervene and try to change their course. In that choice lies a form of freedom, and its value is immense.

1.1.2 Phenomena

Another constitutive feature of scientific knowledge is its high degree of generality. The object of science is indeed, in principle, *absolutely* general. To say it with some philosophical jargon, it includes the entire spectrum of possible *phenomena*, a term of ancient origin that essentially stands for *anything that appears to us in experience.*

The domain of science's analysis is thus potentially ubiquitous (and in some ways infinite) since the family of phenomena it studies is vast: at least in principle, science can be done on anything. Certainly, every scientist will study specific categories of phenomena. For example, the modes of virus propagation or the ability of a certain material to act as an electrical conductor in certain contexts. Nonetheless, the generality and potential universality of a scientific study's content remain.

In brief, science enables us to answer questions of the form: "What happens if E occurs?" where E stands for any event. For example: "If the state of this system is known now, what will its state be in the future at time t?" Or: "If I walk from A to B, knowing the distance between A and B and knowing the number of steps I take in a minute and their length, how long will it take me to walk the route?"

We must not, however, be misled by the prejudice that science deals only with so-called "natural" phenomena. To draw a hard boundary between nature and other spheres of human activity is to misunderstand the reach of the scientific mindset. Consider a question that seems far from fundamental physics. A company wants to know if its sales would increase by making its cookies 10% sweeter and 5% saltier. Suddenly, a host of scientific disciplines come into play. Chemistry is needed to understand the new recipe: nutrition science to evaluate its effects, economics to model the cost and price, and psychology to predict consumer desire. One could even argue for political science, weighing the public health implications.

The object of study is a cookie, but the method of inquiry—forming hypotheses and testing them against evidence—is purely scientific.

In principle, nothing is off-limits. Science can be done on any phenomenon, including the very practice of science itself, provided we find a rigorous way to study it. So, where does this leave us? We can begin to stitch these ideas together. We see that science is an exploration, disciplined by prediction, and aimed at generality. This brings us to an operational characterization, a working definition, somehow self-evident, that summarizes our journey so far:

> Science's task is to describe phenomena, whether current or past, with the ultimate aim of predicting occurrences of future phenomena; to do so, it explores unexpected relationships among phenomena, guided by hypotheses.

1.1.3 Between Practice and Theory

The eminently practical value of science, therefore, lies in offering a key that unlocks countless other activities. It achieves this by allowing us to predict a potentially infinite range of phenomena that have not yet been explored. But how does it perform this remarkable feat of foresight?

At the risk of oversimplifying, we can say that the key lies in the central role of *scientific theories*. Science performs its powerful practical functions not through an accumulation of isolated facts, but through the sophisticated and abstract frameworks that theories provide.

Here, however, we encounter a subtle and fascinating distinction. If we were to ask how to measure the value of a theory, then we would find that its immediate practical use is often not the primary metric. The main goal of a theory is instead to realize certain *epistemic virtues*—qualities that signal it as a solid foundation for knowledge. A theory must, of course, be *accurate*, aligning with observation and experiment. But it should also possess deep *explanatory power*, answering not just *what* happens, but *why*. We value theories that are *simple* or elegant, seeking the most concise explanation for the widest possible range of phenomena. And they must be *internally consistent*, a logical structure free from contradictions.

In other words, a scientific theory need not be useful in a practical sense at first. Its immediate purpose is to constitute a reliable basis for forming beliefs about a part of reality. Its immense utility is a secondary, albeit magnificent, consequence of this relentless pursuit of truth, elegance, and explanatory power. A theory is not built to be a tool; it is built to be *true*, and in its truth lies its incredible power to become a tool.

This pursuit of precision brings us to an unavoidable conclusion. There is, as yet, no known way to express such reliable and virtuous theories without resorting to specific and technical languages. The demand for accuracy and consistency cannot be met by the charming ambiguities and shifting contexts of everyday speech. The language of science, whether expressed in mathematics or specialized terminology, is not an attempt to exclude but a necessary instrument to achieve the required level of rigor. In other words, it is not *any sort* of describing and predicting to be constitutive of the scientific activity; rather, science can take place only within very specific forms of descriptions (and their related predictions), couched in specialistic languages.

Thus, we cannot fully address the question, "What is science?" without first confronting the fundamental role of its language.

1.2 SCIENTIFIC LANGUAGES

1.2.1 Preliminary Considerations

It is possible to become so lost in language that one believes the entire world is a product of our words and their internal logic. This is not a trivial idea; it is a profound philosophical stance. A famous example comes from the Austrian-English philosopher Ludwig Wittgenstein, who in his 1921 *Tractatus Logico-Philosophicus*, propositions 5.6–5.61, declared: "The limits of my language mean the limits of my world. Logic fills the world; the limits of the world are also its limits."

This perspective, where reality itself is bounded by linguistic structures, is a seductive one. But science, by its very nature, must begin from a different assumption: that there is a world external to our language which we seek to describe. If science were to accept the strict linguistic limit, then its quest would not be to understand the cosmos, but merely to perfect its own systems of symbols, losing all connection to the empirical world it seeks to explain. Furthermore, it is also evident that science is shaped as an activity that depends upon the specific material conditions under which it is carried out: the specific characteristics of the scientists' bodies, of the tools they employ, of the ecological and environmental conditions under which the activities are carried forward, etc. For this reason, we must approach the role of language in science with immense caution and recognize its importance, on one hand, while acknowledging its specific theoretical roles, on the other.

To understand the specialized language of science, we must first appreciate the tool it grew from: natural human language. Any language contains expressions that combine into statements, which in turn express propositions or thoughts. Consider the simple statement, "Today is Monday." When a person—let's call her Faffa—utters this on a Monday, her thought aligns with the state of the world, and we call the proposition *true*. If she says it on Tuesday, it is *false*.

This descriptive use, however, is merely one of countless functions. As the field of language pragmatics teaches, the richness of language lies in its flexibility. Faffa could use the same words to ask a question in a moment of uncertainty ("Today is Monday?"), or cry them with an exclamatory tone ("Today is Monday!") upon remembering her car is parked in a no-parking zone. The phrase could be a secret code, meaning she wants mango ice cream, or even a creative act—pointing to a day on a calendar and declaring, "I'll call *this* day Monday."

This subtlety is the magic of natural language. It is a tool that binds communities, expresses emotion, enacts laws, tells stories, forges promises, etc. These functions rely on a vast, invisible scaffolding of shared culture and unstated assumptions. We understand the code, the joke, or the lament because we are insiders to the human context in which it is spoken.

But this very richness, so vital for human society, is a liability for science. The scientific enterprise depends on the creation of stable, universal knowledge. A theory must mean the same thing in Zurich as it does in Tokyo; a prediction cannot be a statement of fact to one researcher and a secret code to another.

Therefore, to build its edifice of knowledge, science performs a great and deliberate simplification. It creates *formal languages* that are designed for a specific, limited set of functions. These technical languages intentionally strip away the ambiguity, emotional freight, and contextual dependence of natural speech. The equals sign in an equation has none of the poetic ambiguity of the word "is" in a line of poetry; the symbol for a carbon atom is not intended to carry any cultural nuances that a community may attach to carbon. Meaning is fixed, universal, and independent of the speaker.

This leads us to the bedrock principle of classical logical theory, which has shaped the language of mathematics and science since the nineteenth century. In these formal systems, a proposition has one primary characteristic above all others: it is either *true* or *false*. All other potential functions are secondary to this binary, truth-evaluable nature. This is not a limitation, but the very source of science's power. It is the unyielding foundation upon which the entire scientific method rests, the operating system that makes a prediction truly falsifiable, an experiment that delivers a clear verdict, building reliable knowledge upon the work of others.

1.2.2 The Value of Formalization

We must now reflect on the importance of formal languages, for they help us untangle the profound divergence between *mathematical truth* and *physical truth*. These are two different, though deeply related, realms of knowledge.

The world of mathematics finds its power in formalization, an exercise in simplification and unity. The great French mathematician Henri Poincaré captured this beautifully when he said, "Mathematics is the art of giving the same name to different things" (1908). He meant that mathematics reveals deep, abstract structures shared by otherwise unconnected objects; the rules of arithmetic that apply to apples in a basket are the same for planets in an orbit. By identifying a single, underlying pattern, we give "the same name" to many different things. This is the spirit of a long-standing philosophical principle known as Occam's razor, which prescribes that entities should not be multiplied without necessity, that is, that we should not posit new entities or explanations when a simpler, more general one will suffice.

Science then takes on the audacious task of using these elegant mathematical structures to describe the physical world. And here, in this act of application, we must choose our words with care. When science uses mathematical rules to build descriptions of reality, we should not speak of their absolute "truth," but rather of their "applicability."

A brief look at the history of science shows why this distinction is vital. Newton's laws of motion, for example, are mathematically pristine and possess enormous, world-changing applicability. For centuries, they were considered true. Yet, we now know from Einstein's work on relativity that they are not universally true; they are a stunningly accurate approximation that breaks down at cosmic scales and extreme speeds. Their applicability is vast, but contingent. This is the fate of all scientific theories: no matter how mathematically sound, they may be revised on an unpredictable timescale.

To speak of applicability instead of truth is not a renunciation. It is the sober acceptance of science's role: to sensibly discuss the material world, using mathematics as a powerful means of exploring the world of formal, abstract truths.

This sheds light on the crucial difference between an equation that exists only in the formal world and one that purports to describe the behavior of an object in the physical world. For the latter, a new burden of proof arises. An assertion made *in cerebro*—humorously, with pen and paper—must be tested. We must verify its applicability through a series of measurements until we have achieved a *psychological certainty* of its correspondence with reality.

This is the role of the experiment: a measurement whose result remains unchanged when repeated under the same circumstances. This demand for reproducibility is the final arbiter, the process by which a beautiful mathematical idea earns its place as a reliable piece of scientific knowledge.

1.3 LANGUAGE AND EPISTEMIC VALUE

1.3.1 Communicability and Verifiability

Scientific languages are important because they are the best vehicles for the knowledge accumulated through scientific research. In other words, those languages are the cornerstones of the epistemic virtues—such as accuracy, simplicity, and explanatory power—that guide scientific research and confer related value upon it. It is worth pausing here to reflect on this.

To express the epistemic value of science, we must first ask ourselves: when is information *scientific*? A mere data collection, to use a term today much in vogue but also elusive, is not science. Nor is data analysis, which is instead a tool for producing empirical observations or organizing data and formulating hypotheses. Data without a theoretical perspective is of little use; it can reveal powerful correlations, but it cannot, on its own, provide the causal explanations that science ultimately seeks. (Incidentally, for a detailed view of how data should be collected and distributed, see the monumental *Frascati Manual* [2015].)

Science requires making predictions based on a theory that allows us both to interpret data and to use it to confirm the theory itself. In other words, scientific expressions must be *communicable* and *verifiable*. However, these are values difficult to refine in detail. Let's give it a try.

Let's start with *communicability*. Simplifying a bit, what we mean is that we are dealing with scientific information if we have not only the results of measurements but also the ability to read them and attribute meaning to them. Additionally, these results will eventually be predictable based on a theory. It should be noted that a mature theory might even predict the limits of its own foresight, suggesting that a system is inherently unpredictable, as in quantum mechanics or chaos theory. This discovery of structured unpredictability is itself a profound piece of communicable information. Therefore, a set of results (of measurements or theoretical arguments) that allows us to make assertions that we cannot verify in practice and ultimately predict does not constitute scientific information.

In short, there is a difference between observation and scientific information from the epistemic value perspective. Scientific expressions are objects of theory used to interpret and predict measurement results: they go well beyond experience, organizing it for the purposes of passing along meaningful information. Moreover, communicability means the transferability of scientific information. For example, Newton's second law, or second principle of dynamics, states that "The change in motion is proportional to the applied force, and occurs along the straight line in which the force is exerted." This can be written as an expression like $F = ma$, or "force equals mass times acceleration"; the latter is probably the most compact form for communicating the second principle among ourselves, a package of immense predictive power.

Let's now move on to *verifiability*. First, data must be trustworthy, in some strong sense of the term, especially if humans were to trust the ability of a computer system to work with data with minimal human intervention. An important effort in this area has been made by Shyue Ping Ong, a professor in the Department of Nanoengineering at the University of California, San Diego, and a strong proponent of what are now called the FAIR principles: "Findable, Accessible, Interoperable, Reusable." These principles refer to the most desirable qualities that data used and discovered during research should possess in order to be "machine actionable" with no or minimal human intervention. Ong believes in the importance of data sharing to replicate experiments and procedures: "A work that is not verifiable cannot be considered science by any standard" (Ong 2023; see also Choudhary et al. 2023).

It should also be noted that we are talking about *verification*, not *understanding*. The latter term implies that there is something deeper than a description, but it is then complicated to specify what this depth consists of. Verification, on the other hand, opens a complex debate on the idea of truth and verifiability. It also suggests that there may be expressions of scientific value verifiable only by machines because they are too difficult for us humans. This leads to a situation—almost paradoxical but possible—with machines verifying mathematical proofs when traditionally we humans did not trust machines. Consider, for example, the proof of the Four Color Theorem discussed by Tymoczko (1979), a controversial result at the time precisely because it used a computer-assisted proof, forcing the community to trust a verification process that no single human could perform. It brings to the forefront a key philosophical question: if no human mind can hold the proof, is the knowledge truly ours?

Now, these observations on communicability and verifiability are highly relevant to how we relate to machines. Science can in principle be verified and communicated not only by humans but also by machines. In the words of Donald Knuth: "Science is what we understand well enough to explain to a computer. Art is everything else" (1974). This is a controversial thesis, certainly, because it seems to dismiss the flashes of intuition and creative leaps required to formulate a novel hypothesis—the "art" within the scientific activity that resists computation. We will discuss it at length. Thus, here is an intuitive but less superficial definition:

> Science is an activity whose task is to describe and predict phenomena through the most compact possible expressions, which are verifiable and communicable by both humans and machines.

The act of describing can be powerful: a sufficiently detailed description of something can potentially serve as a blueprint for its construction. Note, we avoided writing "knowledge," instead using "activity," to visualize something dynamic that, over time, assembles knowledge. We will return to this point.

1.3.2 Compactness

In the classical conception of science, the primary objective is to express true statements, typically in the form of conditional predictions. A fundamental tenet is that scientific communications should consist exclusively of such true statements. This raises two crucial questions: Why is this the case? And what does this requirement for truth actually entail?

At first glance, one might think that science could operate equally well with false propositions and their negations. For instance, to convey that a potato cell's DNA is longer than a human cell's, one could state: "It is not the case that the DNA of a human cell is shorter than that of a potato cell." While logically equivalent, this example highlights that truth alone is not the sole consideration. The challenge lies in finding the most compact and unambiguous way to describe reality. This preference is a practical application of Occam's razor, which we encountered in Section 1.2.2. In science, this translates to a preference for the theory that makes the fewest new assumptions—the most compact one. This reveals that the value of truth is intertwined with the principle of *compactness*: scientific communication should be as simple as possible, meaning it should occupy a minimal amount of space for storage and comprehension.

The concept of compactness has a direct parallel in algorithmic complexity. When designing an algorithm, for example, to find the smallest number in a list, its accuracy is paramount, but so is its efficiency. A desirable algorithm is compact in the sense that it requires the fewest possible steps to arrive at the solution for any given list. The quest for more efficient algorithms mirrors the scientific endeavor to condense vast and complex knowledge. This compression is crucial for several practical reasons. First, compact information is easier to search within an archive and simpler to remember. Second, compactness facilitates the standardization of scientific language, which in turn simplifies communication and collaboration among researchers.

Beyond these practicalities, there are deeper, almost tautological, reasons for prioritizing compactness that seem to define the very nature of science. Scientific research is fundamentally a process of constructing compact theories that can account for a vast amount of experimental data. In this light, science can be viewed as a process of compactification. A successful theory allows us to reconstruct the data without needing to repeat the experiments. Therefore, truth is important, in part, because it facilitates this compactness.

Mendeleev's periodic table is a perfect historical example of this compactification process. Before Mendeleev, chemistry was much like a bestiary—a sprawling list of elements with their seemingly disconnected properties. The periodic table was a monumental act of compactness. It organized the elements based on a simple, underlying principle (atomic number and electron shells) and, more importantly, it transformed a static catalog into a predictive, generative tool. It allowed chemists to dynamically predict the properties of known elements and even the existence and characteristics of then-undiscovered ones. It turned a list into a theory.

At this point, however, we must ask ourselves what compactness is. Is it not a subjective characteristic, dependent on the communicator? The language of science must be understood by both humans and machines, yet their modes of reasoning differ significantly. What is compact for a human may not be for a computer processor. Therefore, to determine compactness, a generally applicable method is needed.

This need for a general, non-subjective method is addressed by the formal concept of *Kolmogorov complexity* (see Li and Vitányi 2013). It defines the compactness of an object, such as a string of data, as the length of the shortest possible computer program that can generate it. A compact scientific theory is, therefore, analogous to a very short program capable of producing a vast amount of experimental data, giving us a rigorous way to measure simplicity.

The mathematician G. H. Hardy famously asserted that "there is no permanent place in the world for ugly mathematics" (Hardy 1941). It is hard to disagree. However, we must be careful not to confuse this belief with the idea that aesthetics is the only criterion in mathematics. Obviously, it is not. Compactness might appear to be an aesthetic judgment, where symbolically elegant formulas are valued within the scientific community. Simplicity and elegance are indeed important. However, the pursuit of the most compact description offers a more profound advantage, which explains its prominence in contemporary science.

This crucial advantage is the ability to *generate information dynamically*, through processes like formal proofs, rather than merely retrieving it from a static database or list. A science that disregards compactness would resemble a Medieval bestiary, a detailed and accurate catalog of natural facts. Compiling such a catalog is incredibly time-consuming, and more importantly, it is difficult to use for explanation and prediction. The act of cataloging can be circumvented by discovering a method to construct an object of knowledge dynamically, starting from more fundamental principles and applying operations. This avoids the need to search through an immense list each time the information is required.

This perspective is especially relevant in our current era, where we are increasingly delegating knowledge to machines. Without a compact theory from which to reconstruct this knowledge, we face significant risks. The knowledge could be lost, diluted by a sea of irrelevant information, or even maliciously altered.

Maps provide a powerful illustration of compactness. A highly detailed map is useful, but a map at a 1:1 scale becomes a useless copy of the territory it is meant to represent. It loses its purpose as a model. This idea is captured in Jorge Luis Borges's fragment, *On Exactitude in Science*, which describes an empire so obsessed with cartography that its cartographers created a 1:1 scale map (Borges 1975).

Compactness and brevity are nearly synonymous with simplicity. But "simple" in this context does not mean "easy to solve." A problem's statement can be remarkably simple, yet its solution can be extraordinarily complex. To cite a famous example in mathematics, consider Fermat's Last Theorem, whose statement is one-line long, but the proof is over 120 pages (Mozzochi 2006).

However, it is crucial to introduce a nuanced counterpoint: the risk of premature compactness. While elegance and simplicity are worthy goals, an excessive focus on them too early in the scientific process can be detrimental. Sometimes, the universe is genuinely complex, and forcing data into a neat, simple theory can lead scientists to ignore or discard crucial evidence that doesn't fit. The messy, "ugly," and less compact early models—like the initial, confusing models of the atom—are often necessary and invaluable stepping stones toward a more elegant and comprehensive theory later on. This reminds us that the path of science is not always a straight line toward simplicity, but often a winding road that must pass through complexity to find a deeper, more meaningful form of it.

1.3.3 Applicability

Let's refine our understanding of "truth," moving beyond the idea that it is the sole, fundamental characteristic of a scientific proposition. Outside the pristine realm of a formal system like mathematics, asserting that a statement is definitively "true" can be problematic. For most scientific disciplines, especially those grappling with complex, dynamic systems, a more pragmatic standard applies. It becomes more useful to speak of a statement's applicability rather than its truth value. We evaluate a scientific model not on its perfect correspondence to reality—a standard it may never meet—but on its *explanatory power* (the ability to coherently account for the phenomena we observe) and its *predictive power* (the ability to accurately forecast future outcomes). For example, a pharmacological model is never "true" in a complete sense, but it is considered valuable if it is *applicable*—that is, if it consistently predicts patient responses. This is because applicability is the essential bridge between abstract knowledge and *concrete human action*. It is what transforms a scientific model from a mere description into a powerful tool for developing new technologies, making informed policy, or creating life-saving medicines.

Thus, *applicability* emerges as a crucial characteristic for a scientific statement, often existing in a dynamic tension with the virtue of compactness. While the ideal is a theory that is both maximally applicable and maximally compact, in practice, a trade-off is often necessary. Increasing a model's applicability to a messy, real-world system may require sacrificing some elegance by adding variables and exceptions that reduce its compactness. Conversely, enforcing strict compactness can limit a model's practical reach in complex situations. Navigating this tension is a central challenge in scientific progress. This is particularly evident in fields like biology, psychology, or economics. The formal language of mathematics cannot to date fully embrace the intricate and emergent phenomena these disciplines study, in part because a truly applicable model would be immensely complex. Whether mathematics will one day develop the tools to fully model such systems remains one of the great open questions at the frontiers of science.

1.3.4 The Uselessness of Science

Of the qualities we have so far attributed to science—compactness, verifiability, communicability, and applicability—each is a peculiar property. They possess a formal and theoretical relevance, yet their value in the fabric of everyday life can be elusive and difficult to grasp. This leads us, following our discussion of the epistemic values that inspire science, to a classic and still timely question: must science, to be considered as such, be useful?

To explore this question through the lens of expert viewpoints, we consulted two distinguished scholars. First, Iordanis Kerenidis, a computer scientist at CNRS and the University of Paris Diderot, invites us to draw a parallel with the state of contemporary art. A classical conception of art, encapsulated in the motto "art for art's sake," saw it as a practice pursued for the sake of beauty alone. In recent decades, however, this view has been largely supplanted. We have come to understand that art is not created in a vacuum; rather, we must consider the position from which it is made, for whom it is intended, and how it functions in the world. Art, particularly socially engaged art, must now address its place within society. It must account for how it seeks to render visible what is unseen, or to distort that which is taken for granted. A social and political dimension has become central to both the creation and appreciation of art, a dimension that extends beyond the traditional metrics of beauty or genius.

Iordanis argues that this profound shift has yet to fully manifest in the scientific sphere. He observes that many scientists today seem content to contribute to a self-referential practice, declining to question the broader relevance of their work—for whom it is done and what impact it might have on the world. This insularity, he suggests, isolates the scientific community, leading its members to interact almost exclusively within narrow sub-disciplines, writing papers for a select few and attending specialized conferences, all while sidestepping the challenges posed by larger societal issues. As Iordanis elaborates:

> I remember when we were PhD students, no one asked us why we were studying a particular quantum algorithm, as long as it was a compelling mathematical problem with a technically challenging solution. It is acceptable if, as scientists, we do not know the precise future applications of our current work. However, we must begin to ask the right questions and engage more broadly with technologists, ethicists, policymakers, and industry leaders. Such interactions, and the practical and social issues they raise, are not valued within the scientific community to the extent they should be. We prefer to remain within our strictly scientific circles, pursuing research that advances our academic careers rather than leveraging science to solve societal problems. This, for me, is the primary issue.

Our second consultation was with the British physicist Sir Peter Knight. When asked about the relationship between science and utility, he responded with a brief and rather insightful anecdote. The episode dates to the mid-nineteenth century and features a conversation between the then British Chancellor of the Exchequer, William Gladstone, and the renowned scientist Michael Faraday. Gladstone, it is said, expressed skepticism regarding the practical value of Faraday's research on electromagnetism. Faraday's reply has become legendary: "One day, sir, you may tax it."

The moral of this story seems clear: even if the utility of scientific discovery is not immediately apparent, it may manifest in the long term, and often in ways of incalculable significance. Science for science's sake, much like art for art's sake, is a principle worth upholding, even in the practice of metascience. It must also be said that utility is not limited to direct applications. There are numerous forms of indirect usefulness. This point is powerfully articulated in the 2022 Aspen Institute report, *In Favor of Pure Science*: "The important thing in science is not so much to obtain new facts, as to discover ways of thinking about them" (p. 12). The report further emphasizes that pure science promotes "moral values such as trust; transparency; integrity; benevolence; competition among ideas and not among people; peace; tolerance; [and] international cooperation" (p. 7). Thus, the value of science extends far beyond its tangible outputs, shaping our very modes of thought and fostering the virtues essential to a thriving global society.

Ultimately, the perspectives of Kerenidis and Knight are not as contradictory as they might first appear. They do not present an ultimatum—a choice between a socially engaged science and a purely curiosity-driven one—but rather describe two essential facets of a single, robust scientific ecosystem. To demand that every scientific endeavor has an immediate, demonstrable application would be to foolishly sever the roots of the tree for the sake of its fruit. The story of Faraday is a timeless reminder that the most profound and transformative applications often spring from research undertaken with no practical goal in sight. This is the foundational principle of "pure science": the pursuit of knowledge for its own sake creates a reservoir of understanding from which future generations will draw in unforeseen ways.

At the same time, Kerenidis's critique serves as a vital corrective to potential complacency. A scientific community that becomes entirely detached from the world it inhabits risks becoming a sterile and isolated priesthood, failing in its implicit social contract. His call is not necessarily for all scientists to become activists or engineers, but for the scientific enterprise as a whole to foster a culture of awareness. It is a call for an end to insularity, urging the community to engage in the great conversations of our time and to remain conscious of its own power and potential impact.

Therefore, the ideal state is not a choice between these two poles, but a dynamic equilibrium. A healthy scientific culture is one that fiercely protects the freedom of foundational, curiosity-driven inquiry while simultaneously encouraging and rewarding a sense of social responsibility. It needs researchers who, like Faraday, can explore the fundamental nature of reality without the burden of immediate utility, and it needs those who, as Kerenidis advocates, can build bridges between that fundamental knowledge and the pressing needs of society. One without the other leaves science either short-sighted or irrelevant. Together, they ensure it remains humanity's most powerful engine for both understanding and progress.

1.4 FORMAL AND APPLIED TRUTHS

1.4.1 Formal Language as a Regulative Ideal

Imagine a world where all scientific propositions possess the certainty and clarity of mathematical theorems—a world where only truth and compactness matter. In such a reality, innovation and discovery could be radically accelerated. The work of biologists, for instance, might shift decisively from the lab bench to the laptop, moving from painstaking *in vitro* experiments to elegant *in silico* simulations. Science would proceed not through the manipulation of biological matter, but through the computational study of its virtual representation.

This is precisely the ambition of the burgeoning field of digital twins, or virtualization. The goal is to construct a high-fidelity computerized replica of a physical object, system, or process—be it a factory, a molecule, a city square, or a human organ. This digital counterpart can then be used to run analyses, test scenarios, and predict outcomes far more rapidly and at a fraction of the cost of physical experimentation. Working with a digital twin allows for the study of a system without altering its behavior through direct interaction, and it permits the effortless modification of initial conditions or the integration of new data as it becomes available. Of course, the challenges are significant, demanding immense computational power and, most critically, the correct formalization—or, in the language of software development, abstraction—of the real-world entity.

Despite these hurdles, new scientific fields are emerging that strive for this mathematical rigor as a foundational step toward virtualization. Systems biology, for example, represents a nascent attempt in this direction. It seeks to formalize complex biological systems, such as protein interaction networks, and study them with a diverse toolkit ranging from statistical analysis to control theory, a discipline traditionally belonging to engineering. However, the goal of comprehensively formalizing the life sciences in mathematical language remains, for now, largely out of reach. This is not necessarily due to a limitation in mathematics itself, but rather the profound difficulty of translating messy, non-numerical, and non-geometric biological realities into the pristine language of symbols and functions. We are far from being able to capture phenomena like human cancer or the complete metabolism of a strawberry plant in a semi-formal language, let alone processes on the vast scale of financial crises. In short, the natural sciences—which maybe the notable exception of physics—still await a satisfactory mathematical translation.

Another compelling example of where formalization and virtualization drive discovery is in the study of multi-agent systems (Wooldridge, 2009). These are computational models that simulate the collective outcomes of interactions between numerous autonomous "agents," whether they be software programs, robots, or even biological organisms. A classic application might model the evolution of a plant population by representing how individual plants interact with each other and their environment. Other models can illuminate complex economic phenomena, such as market bubbles and crashes, by viewing macroeconomic patterns as emergent properties arising from the decentralized decisions of individual firms and banks.

These models depend on a formal description of the agents and the rules governing their interactions. To be effective, they must necessarily compromise with simplification, which can affect the plausibility of the context and, consequently, the reliability of the predictions. Yet, their true power lies in observing the whole. The complex, evolving scenario simulated by the model can reveal macro-level behaviors—emergent properties—that are simply impossible to detect by examining the individual parts in isolation.

Thus, for much of contemporary science, a truly universal formal language remains an ideal, a distant shore. "Give me a lever and a place to stand, and I will move the world," Archimedes is said to have proclaimed. We might, in a similar spirit, declare: "Give us a sufficiently expressive mathematical language, and we will represent the world." But this phrase, like Archimedes's, perhaps expresses a hope more than a reality. It is a powerful vision, but one that, for now, applies only to narrow domains of the vast and complex world science seeks to understand.

This necessary compromise with simplification is not a sign of failure, but rather a profound insight into the nature of science itself. The dream of a perfect, all-encompassing formal language is, in essence, a dream of complete predictability and control—the ultimate expression of the demand for utility. It imagines a world where discovery is a linear process of deduction within a perfectly mapped system. But reality resists such complete mapping. And the domains that are defined by emergence, contingency, and an immense historical complexity that may be irreducible. To acknowledge the limits of our formal tools is therefore to acknowledge that science must be more than just an engineering project. It must also be an exploration.

To insist that all scientific inquiry fit a purely formal structure would be like demanding that all art be reduced to painting. It is an attempt to force a single, rigid language onto a reality that speaks in many tongues, ignoring that some truths are best expressed through sculpture, music, or poetry. The true scientific enterprise reveals itself not in the sterile perfection of a single medium, but in the dynamic interplay between the model and the messy world it seeks to represent.

It lives in the scientist's judgment about which simplifications are acceptable—the equivalent of the painter's choice of palette. It is found in the intuition that recognizes a meaningful anomaly, and in the creativity required to devise new ways of seeing the world that cannot yet be written in formal code. Understanding this gap is not a critique of science, but a celebration of its most human and essential quality.

1.4.2 The Limits of Formal Language (through Machines)

One of the most striking aspects of science is the technicality and artificiality of its languages, often incomprehensible not only to lay readers, but also to most scientists that are not insiders into the specific problem that the language is supposed to shed light on. To what extent can we dispense from the use of a formal language to express scientific knowledge? Is it plausible to think that one day ordinary languages (e.g., English or Tagalog) will be the media through which scientists express themselves?

We discussed the matter with Valeria de Paiva, an expert in category theory and knowledge representations at the Topos Institute in Berkeley. The first feature that Valeria brought to our attention is the scale of the challenge:

> A difficult problem to solve is developing methods to efficiently examine the entire scientific literature, particularly in mathematics. Consider that, in 2017 alone, over 120,000 articles were published in mathematics alone.

Just to reinforce this message: by 2019, the Mathematical Reviews Database already contained over 3.6 million articles (Dunne 2019). According to Valeria, the importance of formalizing scientific results must, of course, be recognized. At the same time, if we could express those results using natural languages (like English, which we are using here), then it would be extremely advantageous for knowledge sharing and enhancing the use of machines in the scientific enterprise. The latter is a point that should not be overlooked today, given the recent acceleration in research and applications of new paradigms in artificial intelligence, such as foundation models.

A quite obvious difficulty lies in the implicit content of natural language, an aspect of human communication that machine learning methods do not always correctly reproduce today. Machines are not (as of today) as successful as people in interpreting a conversation between scientists, for example, between two mathematicians, as they cannot adequately interpret nuances of meaning, vague terms, or ambiguous ones.

In this sense, *hyperformalization*, that is, the formalization of every minute detail and nuance of meaning, is advantageous for the use of machines; with the problem that it is, indeed, highly limited. Consequently, hyperformalization seems necessary for operating through machines, but it is inconvenient for people (in addition to being laborious, as it requires a lot of time). If we could teach machines how to correctly interpret the language used by humans in science, then we would probably speed up the mechanisms of scientific discovery. For this, it is necessary to design computer tools capable of translating natural language into a semi-formal language, one that remains readable by humans and, at the same time, processable by machines. In archival science, the term "digitization" is used to describe the transition from paper to digital documents. Here, we are talking about a further step: a type of automatic formalization capable of capturing language nuances.

An example from a quotidian situation may help illustrate the challenge at hand. With the phrase "She managed to close the door," we may be referring to how there was previously an obstacle in closing a specific door; by instant deduction, we humans infer that the door is now closed unless something happened afterward. The phrase suggests implications without asking us to discover them: the statement "She managed to close the door" implies "She closed the door," but it does not necessarily mean that, at this precise moment, the door is closed.

These sorts of challenges arise when we try to make use of scientific languages as well. Alex Kontorovich, a professor at Rutgers University, brought to our attention the difficulty of achieving effective and reliable communication in mathematics. On one hand, we understand the logical description of a mathematical paper because its interpretation is fixed by well-known standards; but on the other, by having a wider range of plausible interpretations for the rest of the text, we risk distorting it and rendering it unusable. Thus, even a microscopic error in stating a problem can lead to proving something undesirable, creating confusion not only for humans but also for machines. Without accurate text annotation, it is possible that the wrong interpretation will be accepted, generating a type of misunderstanding that grows and accumulates as the scientific text is read. In short, compromises must be made, and the right balance must be found between the precision needed by a computer and the simplicity and informality preferred by human minds.

We find Valeria's approach to the challenge of achieving effective and reliable communication in mathematics quite brilliant:

> My goal is to try to overcome this lack not of data, but of annotations on data when working in mathematics, so that we can explore the scientific literature and understand texts in depth. And we should try to extract as much information as possible with natural language before proceeding with formalization.

The path to the success of this union between formal and natural passes through semi-automatic processes such as terminological extraction to find relevant terms in a given corpus of documents, like all the names of mathematical objects or certain key phrases. Identifying terms and concepts through natural language processing is a necessary process, as is a smooth transition from natural language to formal language and vice versa. This transition must be facilitated, because, as Valeria concludes:

> When we collaborate, we speak in our language. Moreover, when you decide to formalize something, you have to make a technical choice and adopt a tool that, all of a sudden, might become obsolete as soon as a better formalization is discovered. See programming languages. However, it is unlikely that the use of natural language will become obsolete: it is an excellent tool that has evolved over thousands of years. I mean to say that we need to facilitate conversations between human mathematicians and between humans and machines. Formal languages are not an adequate answer for this.

These words directly highlight the value of natural language in scientific discourse. Reflecting on them also emphasizes their importance in opening that discourse to a potentially broader group of people beyond narrow circles of specialists.

We elaborated on how communication in science works with Leonardo Quattrucci, who, in his daily work in science policy, mediates between people working in science or engineering and others who use (or intend to use) that science or engineering within their public or private organizations.

> Language is often a decisive tool in interactions between people in science and engineering within a workgroup. When a conflict

> arises, it typically revolves around the use of a term to characterize a process, a result, an object. When I try to mediate in these disputes, I often recall an interview with Ted Chiang, the famous science fiction writer, by Ezra Klein, in which Chiang draws a parallel between science and religion. A certain way of narrating the history of modern science is to see it in opposition to religion; but, as Chiang reminds us, many people of science (even the champions of modern science, like Isaac Newton) were deeply religious. And they shared with religious people the idea that truth was not for everyone, but for those who could complete a certain journey. In short: science and religion might be, from a certain point of view, closer than science and technology. As Habermas reminds us in *The Theory of Communicative Action*, in science, we have a language that reflects the institutions from which it emanates: those institutions are constitutively elitist, as scientific research is not a field for everyone; so their language is understandable and manageable only by certain people and will never be for the whole society.

We will return to these points later. For now, it is important to note two things: first, the use of formal languages cannot be taken for granted in science, aside from the need for technical vocabulary; second, there is some asymmetry between science and technology, especially in the epistemic values that guide them. Simplifying, we will say that science can never produce results accessible to all people, while technology, in principle, can.

This conclusion, that the language of science creates an inherent barrier to public access, casts the earlier debate on scientific utility in a new light. The insularity of the scientific community, which Iordanis Kerenidis lamented, is not merely a matter of cultural attitude; it is actively constructed and enforced by the very language the community employs. The specialized terminologies and formal structures discussed here (that are extremely useful) are the bricks and mortar of the ivory tower.

From this perspective, the call for science to be more socially engaged is not just a call for a shift in research priorities, but a profound linguistic challenge. How can science address societal problems if the language it uses to define, debate, and solve them is fundamentally inaccessible to the society it purports to serve? A precept of this section is that the work of translation and communication is not an auxiliary task performed after the "real" science is done, but an essential component of any truly useful and socially conscious scientific enterprise.

1.4.3 Psychological Certainty rather than Truth

We have suggested that mathematical language provides, to this day, a crucial tool for carrying out scientific research and, plausibly, for articulating dialogue among humans, between machines, or across the two. It is a tool that brings with it defects, but one that we cannot, even hypothetically, do without at the moment.

As just mentioned, alongside mathematical language, we also find experiments, which not only serve as a universally adopted method of verification but are also capable of suggesting new expressions and thus enriching the language itself.

While all mathematical propositions are fundamentally true or false, experiments provide psychological certainty about a scientific proposition. "Certainty" here indicates the sense of a high degree of conviction in the plausibility of that proposition. As we have said, however, it cannot be a logical-mathematical guarantee.

Psychological certainty occurs when there is both experimental verification (repeating an experiment 10, 100, 1,000, 1 million times, depending on the phenomenon being described, in ways that try to refute it but without success) and syntactic truth, i.e., mathematical proof. In these cases, we use the word truth without risk, since we are not dealing with the physical world but with logic, where, within a formal system, theorems can be proven. Thus, mathematics provides the ultimate guarantee against the psychological certainty of experimental evidence, once an axiomatic context is chosen.

However, one could also conceive of a world, perhaps in the future, where mathematical proofs have a more limited role, at least for human reasoning, for example, a world where machines, through sensors, collect data in some form (perhaps numerical) and then, through a learning process, generate expressions (mathematical or in natural language) that provide (almost) psychological certainty. A specific case might be an automatic procedure, managed by a computer, which, by observing a large number of molecular structures, proposes predictions about new structures of some utility that it has not yet observed. However, it is not certain that the machine will formulate a general theory. Here we are already moving toward the idea of machines as tools with some degree of autonomy in the scientific discovery process, machines that go beyond data collection in a mechanical way. We will return to this in detail in Chapter 4.

The degree of guarantee is based on the quantity and quality of the data collected, the interpretation of that data, and the deduction that follows.

We have a higher guarantee when a huge number of experiments are conducted on a certain proposition and when a mechanical or biological intelligence can deduce—with a certain degree of precision—what happens during the experiment. The deduction will be a syntactic expression, i.e., a piece of language, without a theory supported by mathematical proof.

Gradually, we are thus entering the tangle between mathematics and experiment, between machines and human reasoning. This is the intertwining that is gradually emerging today, stimulated by the growth of the interface between machine learning (the term we adopt to refer to *machine learning*) and various areas of physics, and beyond.

One of the recent central directions of this perspective is the use of numerical techniques as a substitute for closed equations. There are lots of differential equations concerning practical matters (e.g., practical problems that emerge in engineering endeavors) that cannot be solved with exactitude, or whose solution would require an amount of work that cannot be justified by the outcome. An example is the study of the many-body problem for complex systems in quantum mechanics. This is a general challenge encountered when trying to predict with fair degree of accuracy the behavior of multiple small particles, such as a bunch of electrons in a complex molecule. To face the challenge, artificial neural networks have been used to provide approximate solutions, for instance, by reconstructing quantum states from experimental measurements. A recommended initial reading on this topic is the work of Carleo et al. (2019), which also serves as a compendium of the broader context of connections between physics and machine learning.

This emerging paradigm—where psychological certainty derived from machine learning begins to rival, or even substitute for, mathematically proven theory—forces us to confront the fundamental purpose of the scientific enterprise. It pushes the debate between "science for the sake of understanding" and "science for the sake of utility" to its logical and perhaps ultimate conclusion.

The future world this section sketches, one where autonomous machines generate useful predictions without necessarily providing a comprehensible, elegant theory, represents the triumph of an instrumentalist view. It is a world that delivers on the promise of application and results.

This stands in stark contrast to the tradition of a Faraday, where the beauty of discovery lies in uncovering a deep, underlying truth. The question then becomes not only one of epistemic values, but also of practical trust. Can we—and should we—truly rely on what such machines suggest?

Without an underlying theory to explain *why* a proposed cure for cancer works, we lack the ability to predict its long-term side effects, understand its interactions in atypical patients, or explain how to fix it when it unexpectedly fails. We are left with a powerful but brittle prediction, a "black box" whose inner workings are a mystery. The psychological certainty it offers could mask catastrophic flaws that a true theoretical understanding might have foreseen.

In such a future, the very definition of scientific success—and safety—is at stake. The core question becomes: can "psychological certainty" in a machine's output substitute for the robust trust built on causal understanding and formal proof? Or does surrendering our need for an explanation mean accepting a level of inscrutable risk that is fundamentally alien to the scientific enterprise? This is no longer a historical debate but an impending reality shaped by the very tools we are now creating.

1.5 LIMITATIONS

1.5.1 The Language of Mathematics Is Fallible

We have come a long way, and it is worth recapping. After a somewhat vague definition, we have seen that a formal language like mathematics has the power to make information compact and unambiguous. We then also tried to say that, at least for now, not all scientific research should ideally be collected and discussed in such a formal language (without forgetting that science communication and translation is crucial), adding that some research cannot be due to its nature: we thus have, on one hand, the certainty of mathematical proof, which is based on the truth of certain propositions, and on the other hand, the psychological guarantee of scientific experimentation.

Let's delve even further into the question of what science is by considering a new example, borrowed from the relationship between physics and mathematics: Hamiltonians in quantum mechanics. A *Hamiltonian operator* (in concise form, *Hamiltonian*) is a mathematical object corresponding to a physical system's total energy. The *ground state* of the Hamiltonian is defined as the lowest possible energy associated with the system. The *first excited* state is the second lowest possible energy associated with the system. Calculating the difference between the ground state and the first excited state determines important properties of a physical system, with applications in various fields, including material science. And here is the relationship between physical theory and mathematical proof.

Toby Cubitt, David Perez-Garcia, and Michael M. Wolf have shown that, in general, the problem of calculating the difference between the ground state and the first excited state is undecidable (Cubitt et al. 2015). In other words, there are cases where, if we ask a computer to compute what that difference is, the computer will not be able to provide an answer, regardless of the machine's computing power. In practice, the computer will not be able to say Yes or No regarding a proposition submitted to it concerning the difference.

The result indicates that there are objects in the physical world that mathematical language, once fed to a computer, cannot satisfactorily describe. It is an undecidability that directly recalls that famously discussed by Alan Turing and Kurt Gödel (Raatikainen 2022). In essence, these findings act like a high-quality camera with incredible resolution that, despite its power, reveals that certain parts of the landscape are fundamentally blurry, not because of the camera's limitations, but due to the inherent nature of those parts themselves. We gain incredible clarity where mathematics applies, but we must acknowledge where its lens cannot fully resolve the picture.

Thus, limits emerge in the use of mathematics in science. For now, there is no plan B to go beyond these limits, and perhaps no need for such a plan. However, awareness is needed that a mathematical, and therefore computational, answer cannot be provided to every physical problem. Here, "computational" is used in its oldest sense, as things may take a different turn if we consider approaches in machine learning.

How does this awareness translate into the general characterization we give of science? To answer this question, we must return to the idea of compactness in scientific knowledge.

1.5.2 Intractable Problems and Undecidable Problems

Tractable problems are those that computers (more precisely, a specific computational model, such as the Turing machine or logic circuits [Arora and Barak 2009]) can solve in a reasonably human time. Intractable problems, on the other hand, are problems that require too much time or memory. What may be useful to mention here is that if a scientific proposition is difficult, it is likely that it will be so for both humans and machines.

There are many intractable problems. Their existence has important consequences. The difficulty of mathematical proofs is inevitable, given the existence of intractable problems. As is probably clear by now, "intractable" and "undecidable" are not synonyms, as they are on two different planes.

Undecidable means, as mentioned above, that the computer could not provide an answer even if we had infinite physical resources: essentially, we cannot discover the truth or falsity of that proposition, even if it is not only purely mathematical but also has a physical character.

Intractability, on the other hand, always admits an answer, but the time required to provide it may well be a few billion years: it is not a problem of truth or falsity but rather of the time (or memory) needed to solve the problem. It should be noted: this is not related to the elegance or compactness of the solution but simply to the difficulty of solving the problem, quantified by the number of operations a computer must perform to solve it.

Intractable problems exist in every aspect of natural sciences and mathematics. The traveling salesman problem and protein folding are just two examples among many. In the traveling salesman problem, we are given a list of cities, and the problem is to find the shortest route that visits all the cities and returns to the starting point. An instance with 50 cities is already very difficult to solve and requires a very long time.

Mathematical proofs also have complexity associated with listing all possible cases exhaustively. If the number of cases is high, then the proof exhibits such complexity as to make it automatically difficult. Not impossible, perhaps, but tremendously difficult.

The point not to be overlooked is that the complexity of a problem is multifaceted. It depends, first of all, on the problem itself, regardless of the procedures one may be capable of undertaking to solve it. Second, it depends on the *procedures* an agent needs to undergo to solve it. Imagine working on your problem with a computer. (No matter how such a computer is built in the physical world—such as Turing machines, logic circuits, or cellular automata—it will need to perform some procedures to solve the problem. These can be described in a so-called *abstract model* of the computer, a theoretical and simplified construct that describes characteristics without specifying physical details. Now, the more complex the problem, the more sophisticated will be the procedures that its abstract model should be capable of carrying out.) Third, the complexity is also *intrinsic* to the problem itself, regardless of the procedures needed to solve it. Fourthly and finally, the complexity of a problem can also be associated with the *expressiveness* of the language used to formulate it: the more expressive a language, the easier the problem may turn out to be to solve (see, in this regard, the field of *descriptive computational complexity* [Immerman 1999]).

Every algorithm requires a number of steps. This number is determined by the type of mathematical tool used to execute the algorithm. It is always possible to define new tools that condense various steps into a single one. However, algorithmic shortcuts are legitimate only if their computational paradigm, i.e., the abstract machine they call for, can be executed in the physical world. If they cannot possibly be physically executed on a computer, then they remain a mathematical artifact, a form of cheating, from a computational point of view. One could say that, after all, algorithms reflect the problem's complexity once the physical world's theory in which the problem is to be formulated and solved is established.

Thus, we can consider an algorithm as optimal for a problem only in reference to certain physical laws. An interesting example connected to this thesis comes from quantum computation. In the quantum computation paradigm, information contained in physical systems whose behavior is governed by the laws of quantum mechanics is processed. Now, it so happens that some problems, such as searching for an element in a list, have lower complexity in the quantum computation paradigm than in all other classical computational models, i.e., those based on Newtonian mechanics and electromagnetism, for example. This is an interesting fact, demonstrating the value of physical laws even in computation and worth considering in some detail.

Imagine a phone book with n rows where the names are not organized alphabetically (talking about phone books today may sound strange to some readers, as these objects are now obsolete). Our task is to find the number of a specific person in the list. Now, in the world governed by classical physics, we can do nothing but flip through the list and proceed by trial and error. In the worst case, we end up reading n names before finding the one we are interested in. In other words, to be sure of finding the number we care about, we have to calculate some function n times. If the list contains 100 names, then we are forced to read all 100, in the worst case. Counterintuitively, in the quantum world, the problem is solved by calculating the function at most $\sqrt{n}$ times (Grover 1996): in the case of 100 names, we have to read at most 10. And it can be shown, which is extremely interesting, that it is not possible to do better, at least appealing to the physics we know. In other words, there is no physical theory that describes the world as we know it, according to which we can find the name by reading a number of names less than $\sqrt{n}$ times.

Thus, different physical laws allow for computing in different and even more or less advantageous ways. We have also seen above that some problems are impossible, no matter the physical laws, as they require too much time to be solved by a computer in any physical system.

Finally, there may also be an additional circumstance related to the set of all computational classes, i.e., the families of problems determined by the difficulty encountered in solving them. This is because it is likely that, by grouping computational problems with the same complexity, we will end up with an infinite number of classes. For example, how do you describe everything in a world where there are infinite species of animals? Here is another argument for declaring some boundaries within scientific understanding.

One way to interpret what we are saying is this: the mathematical reality is much richer than the physical reality because the latter satisfies strict criteria related not only to what can exist in theory but also to what actually exists, perhaps as one of many possible alternatives.

These last reflections seem to suggest that science cannot provide a complete description of the world (we humbly say "suggest," simply because we have not attempted to formalize the idea of completeness to which we refer). Yet, the predictive power of science and the benefit of its verifiability seem, for now, irreplaceable.

Ultimately, these considerations converge upon what is arguably the most fundamental open question in computer science and mathematics: the P versus NP problem. At its core, the problem asks whether every challenge for which a proposed solution can be rapidly verified is also a challenge that can be rapidly solved. The consequences of its resolution are difficult to overstate. Should it be proven that P equals NP, the very nature of discovery might be reducible to a systematic, mechanical process involving hypothesis testing. Conversely, if P does not equal NP, as is widely suspected, then it would provide a formal foundation for the idea that some problems are intrinsically hard, requiring leaps of ingenuity that no efficient algorithm can replicate. Thus, the P versus NP question is not merely a technical puzzle; it is a deep inquiry into the structure of mathematical truth and the inherent limitations of what we can hope to know through systematic procedure.

1.5.3 Scientific Discovery and the Dynamism of Knowledge

We close our excursus on what science is with some comments on perhaps the most central expression in the conversation and evaluation of science:

scientific discovery. Based on what we have done so far, we can propose an operational definition of this expression:

> A scientific discovery is an expression, i.e., a proposition, of which we were not aware or of which we did not know the truth value and which, based on the content it expresses, has one or both of the following properties: being mathematically true and being verified in the physical world through experiments.

These two properties (being mathematically true and being verified in the physical world through experiments) define science itself as a discipline and the method of doing science. Their role is to establish that the expression contained in a discovery describes a repeatable phenomenon (in the same laboratory and under the same conditions) and reproducible (in other laboratories, under conditions as similar as possible to those in the laboratory where the phenomenon was first observed). We will not dwell on whether a scientific discovery truly reveals a fact independent of the observer, as this is not a decisive issue here. It is instead important to note again how, in an abstract formal system (such as mathematics), the truth value lies solely in the proof; in what is associated with the physical world, however, every conjecture has, in a sense, two levels of truth: the mathematical one and the experimental one, obtained through direct interaction with the physical world. In other words, and as we have seen throughout this chapter, not all expressions with scientific relevance are expressible in a mathematical language. Therefore, not all scientific discoveries will have the property (and perhaps the fortune, at least for us) of being verifiable only by working directly with mathematical discourse, but will rather be verified in the physical world through observations and experiments. The expressions perfectly expressible in a mathematical language, on the other hand, must first be mathematically true to count as scientific discoveries—so the proof will be the first hurdle to overcome, even before conducting experiments in the physical world.

We can finally summarize what has been said about discovery by connecting it with what has been seen about language:

> A scientific discovery is an element of knowledge transmissible through expressions of a language (formal or natural) and of which, until the moment of discovery, we did not have psychological certainty.

As an example, let's look at the concept of speed. Speed, specifically average speed, is how fast something is moving overall. It's calculated by dividing the distance s traveled by an object by the time Δt it took to travel that distance. Here, the symbol Δ (Delta) means "change in," so Δt represents the change in time. In this case, we get the formula: $v = s/\Delta t$ where v (speed) tells us how much distance is covered in a specific time—usually measured in meters per second. This formula is foundational in physics and is commonly taught in schools.

This formula is valuable in science because it's a piece of "transmissible" knowledge: it can be shared and understood universally using the symbolic expression $v = s/\Delta t$. Once we know this formula describes real-life situations accurately, we can use it to create new formulas and ideas. For example, if we know how long something moved (Δt) and how fast it was going (v), then we can calculate how far it traveled (s).

If we recognize this formula as a faithful description of physical reality, then we can use it to build more complex ideas by deploying it as a basis for constructing other syntactically correct expressions. For instance, we might calculate things like: *uniform rectilinear motion*: when an object moves in a straight line at a constant speed; *instantaneous speed*: the speed of an object at a specific moment; *acceleration*: how much an object's speed increases or decreases over time; *distance with constant acceleration*: how far an object travels if it's speeding up or slowing down at a steady rate. As we keep building on this initial formula, we end up with a theory of motion. This theory grows by adding new formulas, understanding the relationships between them, and fine-tuning our ideas based on new observations.

This last example highlights, in an elementary way, the fundamentally dynamic character of scientific knowledge, characterized by the possibility of varying the language we can use to express it, the dynamism of the expressions considered, and how these relate to what they represent. As we will see later, the impact of new technologies on science can be characterized (and measured) precisely in terms of dynamism. Ultimately, the scientific enterprise can be understood as the ongoing interplay between the formal systems of expression we construct and the physical reality they purport to describe. A discovery, in this light, is a privileged moment of contact—a point where one of our propositions demonstrates its value by aligning with an observable, repeatable phenomenon. The dynamism we have discussed is therefore the essential condition for this process, the engine that compels us to constantly revise and expand our expressive capabilities to map the world. It is within this iterative cycle of proposition, verification, and linguistic evolution that the fundamental character of scientific inquiry is to be found.

As early as 2011, Michael Nielsen highlighted the ability generated by the advent of the Internet to give rise to collective research on a global scale that proceeded much faster than that done until then (see *Reinventing Discovery: The New Era of Networked Science* 2011). In Chapter 4, we will examine several more recent examples of how scientific knowledge can be made dynamic at speeds hitherto unattainable. The obstacles to overcome are implementing efficient mechanisms for open, transparent, and inclusive collaboration (where machines have a role) and methods of automatic verification of results. This is where digital technologies will make a difference.

CHAPTER 2

"Technology"

Notes on Technology as the Activity of Building Machines and Procedures to Operate Them

In the previous chapter, we touched on certain fundamental and intuitive ideas about what science is and what characterizes it, including the use of a formal language and an experimental method. By no means do we claim to have exhausted what can be said about science. But exhaustiveness wasn't our goal; rather, we intended to lay down some notes to set up a comparison with technology. Now, we turn to this latter subject, approaching it in a similar manner. As we will see, technology is a very different field from science: practical utility is central to technology, while language and epistemic virtues do not, in general, play a foundational role as they do in science. Technology is also a slippery subject, one for which it is difficult to reach an operational definition. To use a metaphor, technology is like a vast ocean of very heterogeneous practices: from ordinary language to formal procedures, from microscopic tools to gigantic machinery. We will first make a crossing of this ocean and then anchor in the sector that interests us most closely: the relationship that technology has with science.

DOI: 10.1201/9781003676614-3

2.1 IN THE SEA OF TECHNOLOGIES

2.1.1 First Intuitions

In the intuitive characterization we gave of science, we focused on its practical utility in predicting phenomena. But what distinguishes technology? In many ways, we can (still intuitively) consider it a complementary activity to science; if we try to express it as a definition, then we would say that technology is an activity (and the set of its results) whose task is to harness natural phenomena through machines and procedures to operate them, to optimize functions beyond their natural state.

A first aspect of this proto-definition that might have caught the reader's attention is our use of the term *natural*. Simply put, all phenomena that occur, including social ones, are subject to the fundamental laws governing the universe (likely formal) in which they take place. Our purpose in using this term, however, is to illustrate how a stone lying inert on the ground can indeed be lifted using a simple machine that offers mechanical advantage, such as a lever; but without the machine, the stone remains on the ground. Similarly, even an artificial heart, by saving a life, pushes beyond the natural state.

The relationship between technology and science often hinges precisely on the fact that the former helps the latter make discoveries beyond the boundaries of what we could naturally know. Fitting examples of machines that have succeeded in this direction are the telescope and the microscope. The functions, now obvious, that these two types of machines manage to optimize, going far beyond the natural capacities of humans, are viewing very distant and very small objects.

The most distant object we can see with the naked eye is the Andromeda Galaxy, which is about 2.5 million light-years from our planet. This is the maximum permitted by our natural state. In this specific case, the natural state corresponds to the use of our body, under normal conditions, without the aid of any kind of machine. On the other hand, the James Webb Space Telescope can observe objects in deep space over 13 billion light-years away (JWST 2022). The enormity of the difference between the naked eye and the telescope is difficult to grasp. It is easy, however, to describe the function optimized by the telescope: in this case, it is maximized to see as far as possible. Incidentally, this function can be interpreted as a utility function; we can express what the telescope does in terms of how useful it is to the person who has it to achieve a certain goal—in this case, observing celestial phenomena.

Another good example, easy to appreciate, is speed. As of the time of writing, in the spring of 2024, the speed record on foot is 44.72 km/h. The record belongs to sprinter Usain Bolt. It was measured in Berlin during the 2009 World Athletics Championships (as reported by the IAAF World Championships in Athletics Berlin). When the function to be maximized is "ground speed," today both bicycles and cars allow us to beat Bolt's record without surprise. Such vehicles are machines, just as telescopes and microscopes are machines.

Building a bicycle requires various materials and mechanical tools. It also requires procedures (understood as techniques, tactics, and adjustments) to operate the bicycle. When going uphill, you should stay seated in the saddle if you want to consume less oxygen and better manage your breathing rate. When going downhill, it is most efficient to brake before curves. Experienced cyclists tell us that, sometimes, you need to keep pedaling even downhill to avoid stiffening your legs before the next climb.

Thus, the bicycle pushes you to speeds and distances impossible on foot but requires technology both for its construction and to know how to ride it optimally. Going downhill, the bicycle harnesses gravity as a natural phenomenon. However, in general, the bicycle makes use of Newtonian mechanics with a certain level of complexity, and there are entire books on the physics of bicycles. In short, the bicycle pushes beyond the natural state.

We, therefore, have two technologies: one for building the bicycle (our "machine") and one for operating it. Another brief example, if we want to be prolix, is given by medicines: the machine is the pill, and the procedure is the dosage, understood as the process by means of which doses and modes of administration are determined for relevant classes of specific cases.

What emerges from these simple examples, then, is a fundamental dichotomy within the concept of technology itself. On one hand, we have the physical artifact—the machine, the tool, the pill—which is the tangible result of a construction process. On the other hand, we have the associated procedure—the technique, the dosage, the operational knowledge—which represents the methodology for its optimal use. This distinction, however, while useful for analysis, is not absolute. In practice, these two aspects are deeply intertwined, existing in a reciprocal relationship: the design of the machine intrinsically shapes the procedures for its use, while the accumulated experience from applying these procedures frequently informs the design of subsequent, improved machines. It is therefore this dynamic synthesis—this co-evolving relationship between artifact and method—that truly constitutes a complete technology and allows it to effectively push beyond the limits of a given natural state.

2.1.2 Observation and Practical Reason

There are also technologies, usually primitive, not associated with artifacts but directly linked to nature. A classic example is the femur wielded by the hominid in the film *2001: A Space Odyssey*. A bone, a stone, or a piece of wood picked up from the ground can be interpreted as tools. They are natural objects, but we can use them as if they were machines.

Rubbing two pieces of wood together serves to start a fire. A stone is used to crack nuts. Without altering the shape of the stone, the creativity lies in the procedure of cracking the nut: placing the nut on a hard surface, picking up the stone, raising it, and so on. The stone is not a machine in itself. It is an object of the natural world, formed by nature over millions of years, that is used as if it was a machine. Even tools, therefore, become technological objects through our use, through the technique employed to use them. Technology lies precisely in the knowledge of that technique.

Intelligence is needed to define a function that one wants to maximize and to understand how to use the stone to do so. We will call this function, in general, the (utility) function of a technology. Moreover, here, by intelligence, we mean the ability to solve a new problem. We are adopting a superficial definition. Thus, the stone becomes a machine. In these primitive cases, technology lies in assigning to the stone the ability to maximize the function, going beyond what can be done without the stone's use.

Other animal species, besides *Homo sapiens*, also use technology, presumably in a more primitive form. The sea otter picks up a pebble, adjusts it on its chest, and bangs the shell of an unfortunate mollusk against it to open it. It may be an innate behavior. Albert Einstein wrote that "Experience is the alpha and omega of all our knowledge of reality" (1934).

It is precisely by observing the world around us that we can have the spark of intuition about using a piece of wood to do something that previously seemed impossible or that was performed with more expenditure of energy and effort. The otter may perform this action spontaneously because it is already "programmed" in its brain by millions of years of evolution, or it learns it by observing its peers.

The example of the otter leads us to reflect on the relationship between experience and intuition, on the one hand, and technology, on the other. Perhaps the invention of the wheel came to light from the casual observation of a stone rolling down a steep hillside, combining experience and intuition. These two alone, however, do not mean science. Nor do they mean technology.

Technology is born in conjunction with the attempt to solve a problem in the real world. Sometimes it is born as a solution to a problem that we cannot yet formulate, or a problem that does not yet seem to exist. It is not unusual to have solutions already available for which a corresponding problem must be found.

Therefore, technology does not always develop from scientific knowledge. As Foucault proposed,

> As a context, we must understand that there are four major types of these "technologies," each a matrix of practical reason: (1) technologies of production, which permit us to produce, transform, or manipulate things; (2) technologies of sign systems, which permit us to use signs, meanings, symbols, or signification; (3) technologies of power, which determine the conduct of individuals and submit them to certain ends or domination, an objectivizing of the subject; (4) technologies of the self, which permit individuals to effect by their own means or with the help of others a certain number of operations on their own bodies and souls, thoughts, conduct, and way of being, so as to transform themselves in order to attain a certain state of happiness, purity, wisdom, perfection, or immortality.
>
> *(1988: 18)*

This passage gives an idea of the breadth of the spectrum of technologies, even without delving into the proposed partition (for example, one could argue that it would be worth adding a separate category for technologies that serve to improve or make other technologies possible, such as a drill for tightening bolts in building a machine). For our purposes, we can obviously distinguish a *sui generis* branch of technology, namely one that contains only technologies that depend on science. An example is an ordinary electrical wire: a technological invention that depends on knowledge of electromagnetism.

It should also be emphasized, again, that a technological invention arises for something practical: to optimize a (utility) function. Above, we talked about speed. Covering the distance between two places in the shortest possible time is the problem to be solved. While this is the central problem, we can add side problems, such as "using as little fuel as possible" or "in the most comfortable way possible," where the degree of comfort is likely challenging to define precisely, since it is quite a personal matter, or "with the least possible environmental impact."

Ultimately, the presence of these "side problems" illuminates the deeply normative dimension of technology. The very act of choosing which functions to prioritize—comfort over raw speed, environmental impact over cost—is not a purely technical calculation but an expression of human values. Technology, therefore, is never truly *neutral*; it is inevitably shaped by and a reflection of the societal, ethical, and personal priorities of its creators and users. This suggests that any robust definition of technology must account not only for its functional capacity but also for the axiological framework within which it is conceived and deployed.

2.1.3 Formulation and Discovery

The applications of technology require discovery. The function that the invention of the knife originally aimed to optimize is far removed from the preparation of sushi as we understand it today. Indeed, the first knives with obsidian or flint blades were survival objects in prehistory. Their applications were a fraction of those of modern knives. The original function is likely to apply pressure to a narrow surface to cause a fracture in the material in contact with the knife. Over time, specialization grew, leading to double-edged blades, Swiss army knives, Swann-Morton scalpels, and so on.

Specialization means taking the original function and altering it according to the specific purpose, such as cutting different materials.

The discovery could be driven by the search for how to solve a specific problem. Note that we have not used the term invention. It is about optimizing a function that may be generalized or specialized in the future. See the different categories of knives.

The starting point for any technology is, however, a function, such as cutting, heating, lifting, flying, preserving, and pushing. All verbs. All functions. Functions that can also be composed, opening the possibility of trying to build machines that can optimize functions with multiple variables.

Like science, technology also finds it essential to formulate and potentially formalize the problem to be solved. By formulation, we mean a way to describe the problem unambiguously. By formalization, we mean something more technical, often requiring a formal language and, as in science, a language consisting of mathematical terms.

A problem that cannot be quantified—in other words, not measurable—and not formalizable remains unsolved and is probably unsolvable. Based on this, the use of at least partially formal language is indispensable, even when we speak of technology at an operational level.

This aspect of formalization is an essential overlap between science and technology, which often leads to confusion between the two. Both make use of specific languages, technical jargon. Often, technology uses the language of science or, at the very least, its own technical jargon, as it takes scientific or scientifically characterized knowledge as a starting point, where ambiguity is avoided. For example, two tailors collaborating on making a dress will speak a common technical language, as will two plumbers repairing a heating system. From this knowledge, technology generates a practical problem in the world and devises its solution. The description of how to build a machine (or a procedure to operate it) that will lead to the solution is precisely what we feel comfortable defining as "technology."

2.1.4 Serendipitous Discoveries

Not all technology starts from formalized thought. There are also serendipitous discoveries, of which the history of technology offers plenty of examples (see Pievani 2024). One of them is penicillin, probably one of the earliest forms of biotechnology. The author of the discovery was Alexander Fleming, in 1928 (Diggins 1999). Penicillin G for Gram-positive bacteria was discovered after Fleming moved a Petri dish, which he himself had contaminated with the bacterium *Staphylococcus aureus*, to make room on the desk for someone with whom he shared the office. Afterward, Fleming went on vacation, leaving a window open. The culture in the dish was contaminated by a spore that likely entered the room through the window. When Fleming returned from vacation, he noticed mold in the dish, and near the mold, the bacteria had died. When he noticed, he said, "This is funny." As Vincenzo Tiberio wrote beside a similar observation, "the properties of these molds are a strong hindrance to the life and propagation of pathogenic bacteria" (1895: 95, our translation).

The rest is now history. It is relevant to emphasize that what Fleming discovered is a natural phenomenon. Nature provides a fact, something that is verifiable and repeatable under the same conditions, bringing us back to the definition of science proposed in Chapter 1. This fact suggests a way to use something natural with a procedure that requires creativity in inventing the use of a phenomenon, in this case, producing antibiotics to fight bacterial infections in people and animals. As in the example of the stone picked up to crack a nut, the mold observed by Fleming has many different properties, some of which can be applied to solving a problem. They can be applied to maximizing some function.

This is an example of how a discovery, when we talk about technology, can involve discovering a phenomenon and using it in a precise way. The discovery could be a coincidence. But it would be incredibly fortuitous to accidentally discover the Otto cycle engine (the traditional engine that converts chemical energy into mechanical energy) without prior knowledge of thermodynamic or mechanical phenomena, just as it would be incredibly fortuitous to accidentally discover the Charmat method (the one we use to produce Italian sparkling wine) without knowledge of fermentation processes.

It is at this stage that engineering comes into play as a set of techniques for creating technologies: from discovering the phenomenon to the adoption of a machine by a user or customer. Today, in the evolution of technology, it is sometimes considered essential to work from known problems, that is, from problems worth solving because one has an idea of the consequences brought by the solution. This procedure is sometimes called working backward, precisely because it tries to determine the importance of solving a specific problem based on the impact the solution has on a certain group of people.

The connection between the natural phenomenon and the problem is a key point in how technology is born and how technology is a non-scientific activity, at least on the surface. Science and technology remain very close, at least in appearance, but do not share the same final goal. Science ultimately seeks to predict the future of a particular phenomenon based on descriptions. Technology does not aim to predict, explain, or describe but rather to build machines and procedures with specific functions, which can certainly be "predict" and "describe," but only if seen as functions. If science explores the boundaries of human knowledge, then technology has the goal of concretizing results for practical reasons. They thus appear as two different but connected planes.

Yet, this distinction does not imply separation; rather, it suggests a symbiotic relationship. Science provides the fundamental understanding of natural phenomena that technology can then harness for practical ends. In turn, technology provides novel instruments and tools that enable science to make new discoveries. This reciprocal dynamic creates a powerful, self-reinforcing loop, where progress in one plane often becomes the necessary catalyst for advances in the other, driving human capability forward in a way neither could achieve alone.

2.2 UTILITY, UTOPIA, DYSTOPIA

2.2.1 Primary and Secondary Technologies

Let's now delve a little into the articulation of types of technologies. A distinction that will prove useful is that between primary and secondary technologies. A technology is primary when it makes use of a single natural phenomenon or, at least, predominantly a single natural phenomenon. Examples include penicillin and fire. Any technology that uses two or more phenomena is secondary, and one could even go further with this classification, including more than two phenomena.

A primary technology typically relies both on a scientific understanding of the phenomenon, understood as a compact, predictive, verifiable, and communicable description, but, above all, it relies on the ability to control its occurrences. Without the ability to light a fire at will, the technology of using fire as a method of heating a cave would be useless.

The same idea applies to penicillin. Going out to find molds and eating them? It doesn't seem plausible. It is done with medicinal herbs because there is a tangible benefit and, perhaps, because both the complexity of the technology and, simply, the monetary cost of producing the same therapeutic substance in the laboratory do not invite industrial production. Similarly, it doesn't make much sense to produce petroleum in vitro unless this happens once new, deeply innovative industrial processes are developed.

That said, every technology requires some background knowledge and, often, other technologies. It is difficult to keep these components separate in classification. Putting together different chemicals with some purpose in mind is also a primary technology, but it ceases to be so when we need to rely on multiple phenomena. For example, gunpowder is obtained by combining charcoal, sulfur, and potassium nitrate in precise dosages. How are these substances obtained? If they are natural and do not require specific procedures to be made available, then gunpowder is a primary technology.

This demonstrates the artificiality of certain, albeit useful, definitions: the choice of classification of types of gunpowder just proposed would still remain debatable since, for example, mining technologies (or the French method) are needed to access potassium nitrate. The distinction between primary and secondary technologies is, therefore, as elusive as concepts like "control of a phenomenon" and "single phenomenon" can be. To control a phenomenon, it often takes control of many others, in a number that may remain undefined; moreover, for a phenomenon to occur, obviously, others must occur as well.

Another important thing is to distinguish between the context of discovery of a technology and its execution. The crucial question for distinguishing between primary and secondary technologies, then, is: once a machine or a procedure for a technology is developed, how many natural phenomena are involved? Crop rotation as a technique in agronomy is a good example. It could be considered a primary technology because it uses knowledge of a very specific type of phenomenon; yet, historically, crop rotation required many experiments before being adopted.

We should not confuse the classification of primary and secondary technologies with the complexity of discovery, the need for basic knowledge, and access to other technologies. There are many techniques for lighting a fire, and there are many different fuels: optimizing techniques and fuels has an important effect on technology but does not alter the fundamental natural phenomena that make it possible.

We then come to secondary technologies. An electric car uses a motor that converts electrical energy into mechanical energy through a rotating shaft. It thus uses phenomena related to electromagnetism. It also exploits friction in the brakes to stop at red lights. It uses the wheel. It uses batteries. These are very different phenomena: for this reason, we have a secondary technology.

In ridesharing services, the passenger is automatically paired with a driver through an app on a mobile phone. Ridesharing is a secondary technology because it employs a large number of natural phenomena: electromagnetism related to the use of the phone, a satellite positioning service, and cars. Other examples include television and the semantic network (World Wide Web). In Tim Berners-Lee's dream, the inventor of the Web, it would help reduce bureaucracy, making it more feasible that "our daily lives [are] managed by machines that talk to machines" (1999). Indeed, the Web is nothing more than an information container, largely readable by machines: it integrates, among others, the technologies associated with the machines themselves and the technologies for putting machines in communication with each other.

This integrative nature is characteristic of most modern, advanced technologies. The power of a secondary technology often lies not merely in the sum of its constituent parts, but in the novel capabilities that emerge from their combination. As more primary and secondary technologies are layered upon one another, the potential for new functions grows. This layering effect is precisely what allows a relatively simple device like a mobile phone to become a gateway to complex systems for navigation, communication, and commerce, far exceeding the utility of any of its individual components. And an important consequence of this layering is a principle of *abstraction*. For a complex secondary technology to be viable, the underlying technologies it integrates must be treated as reliable "black boxes." The driver of an electric car does not need to understand the solid-state physics of its battery, nor does a web user need to grasp the protocols of packet-switching. Each layer of technology encapsulates the complexity beneath it, presenting a simple interface to the layer above. This functional abstraction is what makes the construction and use of immensely complex systems possible, as it allows engineers and users alike to operate at a high level without being overwhelmed by the intricate details of every constituent phenomenon.

2.2.2 Five Parameters

The definition we have given of technology is not based on its utility. There are obviously countless perfectly useless technologies that were once extremely useful. In *The Adventures of Pinocchio*, Carlo Collodi explains that Lampwick was called that "because of his slim, trim figure—skinny and lanky—just like the new wick in a little night lamp" (2021: 190). Well, today many people wouldn't even know what that wick looked like because it's a useless technology in the contemporary world. But that doesn't take away from the fact that it is, nonetheless, a technology. A technology can, therefore, be useful at one time and then become useless; and, from useless, it can become useful again. In an emergency (e.g., in the absence of electricity), an old radio we keep as a trinket could become very useful again, as could a candle. This is how utility seems to depend, at least in part, on context.

To fully reflect on the idea of utility, let's take a step back and return to the idea of the natural state. The intuitive image of a natural state is a photograph of the world without the intervention of machines built by humans.

Restoring the quality of water in a river seems to restore to its natural state which was degraded precisely by human intervention, even if the process was likely unintentional. The same goes for melting glaciers. In this case, pollution is not a function we wanted to maximize but rather a side effect of some other activity. *Natural state*, then, refers to the photograph of what can be defined as authentic or original, at a specific moment. Technology modifies that photograph by building machines and procedures that alter it; it creates a difference that could not be achieved except through the technology itself.

Again, however, the boundaries are blurred. It is clear that when we speak of the natural world, we are not leaving out humans and societies as part of that world. Human language itself can be considered a secondary technology, composed of the various technologies we use to communicate and the various things we can do when we communicate (e.g., describing, promising, exclaiming, singing, and naming).

In short, it is not easy to distinguish between the state of nature without any technology, a state in which we have technology, and other, paradoxical situations where technology is used precisely to bring us back to the state of nature. The natural state is essentially nothing more than a useful metaphor to emphasize the difference made by a technology compared to a state of the world in which it did not exist. Every technology adds a machine or procedure that alters our ability to take advantage of a phenomenon. With this addition, in the gap between before and after, we can evaluate utility.

The difficulty lies in defining the utility function and, with it, what is useful. In the reconstruction we propose, based essentially on experience and reasoning, this utility function depends on at least five factors: the definition of the function itself; scalability; the context of reproduction; the context of use; and the narrative. Let's look at them in order.

Technologies apply to very different phenomena to perform very different functions. We have social technologies, codified in norms; for example, how to behave in line to order coffee in a coffee shop. There are culinary technologies, perhaps explaining how to cook pici with garlic: a procedure involving stoves, pans, ladles, knives, and ingredients. And so on. The technologies we are primarily interested in here are "scientific" ones, that is, those based on scientific knowledge of one or more phenomena.

Regarding scalability, it should be noted that not all technologies are equally scalable. A few years ago, physicist Barbaros Özyilmaz told us about the outstanding ability of a student to spread graphite on a sheet of white paper (with a common pencil) to study the electrical conduction of certain nanoparticles. Although, in principle, the technology was known to everyone in the lab, the experiments required the student's presence.

Another classic example is cooking recipes, whose execution depends on the hand of the person performing them. Some recipes can be carried out by many people, but not by everyone. For example, performing the cut of a Florentine steak is not for everyone, just as rolling out pizza dough according to the rules of the art is not for everyone. There are also personal and perhaps inimitable skills. The cuts of some sushi masters are legendary, but even in the family sphere, we have personalized skills to which we become attached: Andrea's grandmother Annita cut potatoes for frying in a way that no one else in the house knew or has known how to imitate satisfactorily.

The more a technology depends on a single person or certain characteristics that only some people possess, the less scalable it will be. The assumption is that a scientific technology can typically always be performed by a machine and, therefore, is in principle scalable *ad libitum*. In theory, the information acquired to solve practical problems, such as, for example, the maneuvering of a sailboat docking, could still be learned and thus scalable, once the necessary information is precisely determined.

But, of course, returning to talking about recipes, a theoretically scalable recipe may not be so in practice. It depends on the context of reproduction. Truffle tagliolini are exquisite, and the recipe is known, but we can only cook them when we have the raw material available. To give another example, batteries for electric cars depend on the availability of lithium. In addition to raw materials, to perform a technology's function, we also need the machines, the production sites, and so on. Each technology will have its requirements, sometimes very specific.

Not everything we produce will be useful to everyone and in every context. Technology, therefore, also depends on a context of use. There may be scalable technologies with a well-established reproduction context but a limited context of use. For example, a microprocessor built to solve a particular type of computational problem may not succeed in the market because the community of developers capable of using it is too small.

Sometimes, moreover, a machine built to perform a specific function can be employed in a new context of use. Simply put, a hammer has various applications. It serves to drive nails. It can be used to apply pressure to something. When assembling a dining room table, I use the hammer to make wooden joints fit together. Surely, the hammer was conceived to solve a problem we don't know, but today it serves multiple functions. It is, in short, a machine that appears in many different contexts.

Now, the utility of a technology concerning a context of use can be seen as maximal, or less so, with respect to established parameters such as cost, scalability, or reliability of the function. Staying with the hammer, since the physics of the hammer is clear, it is not difficult to provide a formal abstraction (through the language of mathematics) of each of the functions the hammer can maximize. The fact remains that to describe a technology in which the hammer appears, a problem defined by the person using the hammer is always necessary.

There is no such thing as a useless technology, based on the very definition we have given of technology. A technology is superseded by a better one for the function to be optimized only when the function has been defined.

The relationship of compatibility between functions can also be very relevant and delicate to manage. This typically happens when, in trying to maximize one function, we end up minimizing another. We are faced with a complicated problem of optimizing correlated functions, the solution of which requires setting priorities: in certain cases, it is better to set functions not to be altered (too much) in the necessary compromise of optimizing a wide range of functions.

We can summarize this point about function compatibility with the slogan "you can't get something for nothing," not even in technology. An extremely important and well-known example of this is related to the environment. If a technology produces a demonstrably polluting substance but is useful for the function f, then the relationship between benefit and harm in the process of maximizing f must be considered. The use of the substance must be reconsidered, and alternatives must be determined. This is one of the reasons it is essential to form ethics committees composed of people with heterogeneous views and skills, who can assess the spectrum of all the effects of a particular technology. This is because it is likely that some of these effects will go unnoticed until they have a self-evident negative impact. Sometimes, the damage is done, and it has become too difficult to repair it.

There is also, and this is sadly common, the wrong application of technologies. A hammer cannot be used as a microscope and should not be used as a weapon. We can also, paradoxically, conceive of a 50 kg fountain pen. We can conceive of a shoe worn over another shoe, which in turn can be worn over yet another shoe. What function are we maximizing? It is not trivial to find examples, but there probably are technologies that do exactly that, and we invite the reader to think about them.

There are, in fact, many examples at the boundary between technology and art. Bruno Munari built useless machines as works of art: "useless because they do not produce, do not eliminate labor, do not save time and money, do not produce anything marketable" (1937, our translation).

Following the definition at the beginning of this chapter, a machine is useless if it does not allow us to do something beyond the natural state. Based on this, Munari's phrase is much more specific, as it associates the concept of productivity in the industrial context with machines, somewhat expressing the *zeitgeist* of the first half of the twentieth century.

Is a machine as an art object useful? We do not venture into discussing this profound topic, also because it would take us far from the purpose of this chapter. However, it seems to us that a work of art does not try to maximize a function beyond a definable natural state, at least at first glance. The work of art can serve as a means of communication. However, the message communicated is not subject to a truth value—a term borrowed from logic. The message is not classifiable as true or false. In the case of art, we have a form of representation open to interpretation, unlike science, which describes the natural world.

Of course, even art is subject to the laws of physics. What cannot be imagined and represented does not seem to belong to the sphere of artistic objects. From a different angle, art is an activity that connects the imaginable world with the representable world—art as a tool or process for representing thought or felt things. The observer, the art consumer, has their attention captured or not based on what the art object evokes.

At the moment when what is created attempts to maximize a function beyond the natural state, we can therefore speak of technology. Art and technology appear thus connected, as do art and science. Art imagines and realizes. Science describes and predicts. Technology imagines and realizes with the aim of solving a problem.

Technology and art intersect further when we talk about design. This is obvious. The aesthetic side has value, although presumably subject to or dependent on a specific culture. This value can be seen as an additional function to be maximized in the context of use, imposing restrictions on reproducibility. Design plays with aesthetics in machine building, generated as it is by the desire to combine art with functionality. Design as an art form thus helps the adoption of technology and can, retroactively, influence the technology itself. Again, technology, as a bridge between the imaginable and now buildable world, requires a specific practical function; otherwise, it becomes a way to collect useless machines in Munari's sense.

We then come to the fifth and final characteristic: narrative. It is worth dedicating a separate section to this because today, narrative plays a key role in the process of developing and adopting technology. Moreover, narrative, seen as a marketing tool under the specific name of storytelling, is increasingly permeating the context of scientific research as well. Thus, new problems arise, where communication-related aspects are crucial in pushing one technology over another. And communication has nothing to do with the quality of a product or its value. Indeed, in ethical discussions around technology, the general attitude tends to downplay narrative and focus on principles or, potentially, impact, when it is precisely the narrative that is one of the reasons behind the success of new technology, at least on a limited timescale.

2.2.3 Storytelling and Technology

It is undoubtedly an illusion that every technology in our society is invented to solve a useful problem. The definition we gave above is abstract; it does not differentiate between technologies based on their value, whatever type it may be. Going beyond the definition, the development of technology and the push for its adoption can also be interpreted as a means of increasing wealth, regardless of the benefit it brings and its ethical status, broadly speaking. And it is difficult to define benefit and ethical status: one enters the labyrinth of moral philosophy and must devise appropriate strategies to involve a sufficient number of people representing sufficiently diverse perspectives.

It is evident how technology, once associated with the everyday, can be driven to—and thought of as—solving certain problems, formulated with the primary intent of generating profit. Hence, the need to identify a market niche and determine which specific problems a technology can solve for a particular type of consumer. Here, a scene from Season 2, Episode 1 of the TV series *Silicon Valley* comes to mind. Gavin Belson, the CEO of a large technology company, says: "I don't want to live in a world where someone else makes the world a better place better than we do."

A recurring theme in the series is ultra-specific and incomprehensible technology, advertised as something revolutionary and important to make the world a better place. In the series, the philanthropic side hides a rather shoddy marketing strategy. It is almost necessary to adapt to the trend. In the world of *Silicon Valley*, companies must always talk about "making the world a better place" to appeal to customers and investors.

The important point here is that technology needs a story to tell. In this way, it differs from science (at least traditionally). Science, technology, and art all revolve, in some way, around the material world. As mentioned above, the first to predict the world; the second to modify the world; the third, to represent something often impossible to formalize and interpret.

We can then summarize what has been said so far as follows. Science tries to tell a story that speaks of the world based on information acquired from observing the world and manipulating symbols; it tries to interpret the world, translating it into our language. Technology tries to tell a story while building machines compatible with the story being told, where sometimes the story precedes the machines.

Intuitively, invention comes before adoption. A startup's deck can sometimes precede the invention itself: this can happen when the idea is in its embryonic state, but also when it is potentially unachievable. Incidentally, the deck is a presentation that the startup uses to raise capital and convince investors or customers; it usually includes information on market size, the product, the use of funds, the relevant people, and the management plan.

In the process of birthing a technology today, the choice to build something can (and perhaps should) be suggested by a specific problem of a group of customers. The possibility of adoption drives the invention, as does the possibility of creating a benefit for those who will use it.

On one side is the invention, independent of adoption. On the other, the analysis of potential adoption suggests what needs to be invented and built. The function to be maximized then becomes governed by, or related to, the value derived from adoption; it distances itself from modifying the natural state, as is obvious, but must be used as a criterion to acquire customers. It can also be entirely revised as innovation and adoption are balanced. In the most extreme case, we are faced with entities whose only goal is to achieve significant adoption, without any benefit to customers.

There are thus cases where the narrative accompanying technological development becomes dominant as a representation of the technology in the eyes of the customer, where "customer" is understood as something very generic, such as society itself. Ergo, an idea associated with technology, becomes, for a company, the engine for creating trust even before the technology itself is developed. This is why, as part of the narrative, phenomena such as celebrity worship or the cult of personality appear, where a particularly charismatic figure contributes to creating enthusiasm for a certain type of technology, even when it is immature. Furthermore, situations may arise where technology becomes an illusory tool for solving artificial problems. Situations can also occur where technology acquires the nature of a status symbol. So you hear, "if your company A doesn't have an AI plan today, you're out of the game," even if there's no evidence that AI creates value for company A's customers. What matters is the company's image, and this image can drive up A's stock price regardless of its actual economic status.

One evening, with a couple of friends, we entertained the idea of establishing an imaginary company called *theplace-b*. The company produces and sells pills to be ingested like so many medicines on the market today, but with a specific feature: the pills are empty. They are devoid of active ingredients and free of ingredients: *theplace-b* certifies the emptiness of the pills, declaring: "With the work of our laboratories, we have put absolutely nothing in the pills." Does the idea have something artistic about it? Perhaps. Are the empty pills new technology? Not yet. Why would anyone buy them? For purely aesthetic reasons, they are perfectly white.

But now let's think about an advertising campaign where we say the following:

> *theplace-b* pills contain absolutely no substance. Take *theplace-b* before an important job presentation: we believe this way, you'll have a chance of making an excellent presentation. Take a *theplace-b* if you're stressed: we believe you'll have a chance of relaxing. You're in the gym and want to feel stronger? Take *theplace-b*, and we think you'll have a chance of being stronger.

There's no fraud in these statements. Why? Because they contain nothing false. With or without *theplace-b*, the above events can happen with a probability greater than zero. So, we wonder whether *theplace-b* is a useless technology. Are we sure? It's not out of the question that *theplace-b* has an effect on the psychology of the person taking it. If each pill of *theplace-b* was expensive, then it's likely that it could have an even greater effect. Yet, we have clearly stated that *theplace-b* is explicitly a placebo.

Thus, technology needs a story within a certain frame of reference. It needs a function to maximize, to be called technology. It needs a story to define its function. The narrative, the story that envelops a particular technology, truly has the potential to impact the world, as it can catalyze the adoption of technology and the approach of those who use it. The narrative, the story, contributes to the acceptance of new technology as much as it does to the assignment of a moral value.

Through the story, an object becomes good or bad; it can also acquire both adjectives one at a time in different historical moments. The change can be determined by events and the modification of the story in a new sociocultural context. Clearly, the narrative does not always generate ethical status: some technologies are repugnant in any case, no matter what is said about them; others are precious to humanity even if only a few people fully understand them. The relationship between narrative and ethical status of technology is complex and delicate: reflecting on it is a necessary step to deliberate who should manage communication technology and how; we will return to this.

Indeed, while narrative holds a unique and powerful place, its influence is best understood in relation to the other factors we have outlined in the previous section. The definition of a function is meaningless without considering its context of use; scalability is constrained by the context of reproduction; and the success of the entire enterprise can ultimately hinge on its narrative. A holistic assessment of any technology requires navigating this multi-dimensional space, recognizing that its true utility is not an intrinsic property but an emergent outcome of the dynamic interplay between its design, its material constraints, its human context, and the stories we tell about it.

2.3 SCIENCE AND TECHNOLOGY

At the conclusion of these first two chapters, it's worth reflecting on the general relationship between science and technology.

It's difficult to talk about science without thinking about technology. An online search for "science and technology" returns a series of quotes from famous scientists reminding us, or feeding us, the solidity of science and the ephemeral nature of technology, such as:

> Science makes technology possible, but technology makes science visible.
>
> – *Adam Osborne*

> Science is nothing without technology.
>
> – *Wernher von Braun*

> Technology is just a machine, science is the intelligence that drives it.
>
> – *Louis Pasteur*

> Science and technology are the lifeblood of any modern society.
>
> – *Jawaharlal Nehru*

> Technology is nothing without the science behind it.
>
> – *Werner Heisenberg*

> Technology is just a tool, and science is the driving force behind it.
>
> – *Viktor E. Frankl*

> We live in a society exquisitely dependent on science and technology, in which hardly anyone knows anything about science and technology.
>
> – *Carl Sagan*

These quotes (or alleged ones), which use disparate metaphors, also testify to the fact that the relationship between science and technology does not seem to be at the center of debates in the philosophy of science, nor the focus of the narratives of science itself.

We can hardly hope to find help in setting up the relationship between science and technology in the public sphere. Rather, we shall resort to a quote by Umberto Eco, from an article in *L'Espresso* in 2004 titled "Credere alla Tradizione" (Believing in Tradition):

> The mass media often put science on trial, holding it responsible for the Luciferian pride with which humanity proceeds toward its possible destruction, and in doing so, they evidently confuse science with technology. It is not science that is responsible for nuclear weapons, the ozone hole, the melting of the ice, and so on: science, if anything, is still the one capable of warning us of the risks we run when, perhaps using its principles, we rely on irresponsible technologies. But in the condemnations that are often heard or read about the ideologies of progress (or the so-called spirit of the Enlightenment), the spirit of science is often identified with that of certain idealistic philosophies of the nineteenth century, according to which history always proceeds toward the better and toward the triumphant realization of itself, of the Spirit, or some other propulsive engine that always marches toward Optimal Ends. And after all, how many (at least of my generation) remained always dubious reading idealistic philosophy manuals, from which it emerged that every thinker who came later had understood better (or 'veritably') the little discovered by those who came before (as if to say that Aristotle was more intelligent than Plato)? It is towards this conception of History that Leopardi lashed out when he ironically spoke of the 'magnificent and progressive destinies'.
>
> *(2004, our translation)*

Eco's words hint at an intricated relationship between science and technology, whose complexity depends also on their (perceived) impact on society. In bringing this chapter to a close, we shall therefore briefly elaborate on both complexity and impact.

2.3.1 A Complex Relationship

One cannot say that science was born before technology. On the contrary, one might assume that archaic proto-technological solutions were introduced by our most remote ancestors, including those not of our species, and who did not have the intellectual capacity to fully develop scientific knowledge. Picking up a stone and using it as a hammer does not require science except in a marginal sense of the term; empirical observation is sufficient.

Nor can we think of starting with science because it is more important than technology. In fact, technology generates today a direct economic impact much greater than science (at least in the short term), also because it has an easily traceable impact on our lives. But here it is not only worth highlighting the term "today": it must also be considered that science generates value, often indirectly. To quote Einstein:

> There are two ways in which science affects human affairs. The first is familiar to everyone: Directly, and to an even greater extent indirectly, science produces aids that have completely transformed human existence. The second way is educational in character. Although it may appear less obvious to cursory examination, it is no less incisive than the first.
>
> *(1935–1936: 13)*

It is indeed clear that the theory of relativity or quantum physics has had a gigantic impact on our mental models and culture, but also on our economy: in a recent study conducted by the scientific organization Institute of Physics in the United Kingdom, for example, it is argued that in 2019, physics generated 11% of the country's gross domestic product (Youngman 2022). While recognizing that it is extremely difficult to isolate and quantify the economic impact of a particular field of science (because there are deep connections between different fields, because it depends on complex social dynamics and is generated over potentially many years, etc.), it remains undeniable that science also has a very important economic value. Greater or lesser than that of technology? Difficult to prove.

So? Why give primacy to science? An interesting hint to answer this is offered by Edward Teller's presumption, who, in his controversial essay on the legacy of Hiroshima, writes: "Today's science is tomorrow's technology" (1962). This is perhaps too apodictic a statement, which we feel like welcoming in the following attenuated variant: "Today's science will be useful in tomorrow's technology." In short, perhaps it is worth starting with science because, as things have worked in recent decades (and perhaps for the past two or three centuries), scientific knowledge influences (temporally and logically) technological knowledge.

But we must not be carried away by the fury of finding a winner: is it not also true that science is influenced by technology? Certainly, some sectors of science seem able to exist and even progress using only a minimum of technology. For example, in mathematics, one could work only with paper and pen, perhaps with the help of printed documents. However, not only do those who do research in mathematics, in fact, use electronic scientific journals and books as technological objects, even the use of paper and pen can also be seen as the use of something technological. The latter, in fact, increases our ability to articulate thought and extend our memory. And here, incidentally, we are still not considering our opinion, which we will develop in Chapter 4, that in the future, mathematics will naturally be the product of interaction between humans and machines much more than it has been so far.

One might at this point retort that paper and pen are nonetheless technologies of presumably lesser complexity than a particle accelerator, an electron microscope, or a computer capable of formulating scientific hypotheses and verifying the correctness of a mathematical proof. Will mathematics in the future be able to survive without complex technology? Here is a question that is difficult to answer.

Already today, symbolic and numerical computing environments are often a daily tool for physicists and mathematicians—a topic we will touch on later in this book—and machine learning is emerging forcefully. It must also be said, then, that many fields of research require much more, beyond paper and pen: think of molecular biology, medicine, high-energy physics, and, in short, everything that requires experimental verification and measurement—which seems to encompass a bit of all science itself, almost by definition.

In short, the relationship between science and technology is complex and cannot be taken for granted in any direction or articulation. Perhaps it should not even be taken in an absolute sense, as they are practices that have evolved throughout human history, intertwining and influencing each other. Not only does science often develop in tandem with technology (think of astronomy and lens technology for optical telescopes), but also the perspective risks being completely reversed: as we will see better later, and as is intuitive, today, it is often technology, armed with its economic and sociopolitical power, that dictates which scientific research lines should be pursued, with potentially dangerous consequences for science and civil society.

This inversion of influence has profound implications for the very nature of scientific inquiry. When research agendas are dictated by technological utility and market demands, there is a risk of systematically prioritizing applied science—research aimed at solving specific, pre-defined problems—over basic, curiosity-driven science. While this may accelerate short-term innovation, it could paradoxically constrict our long-term epistemic horizons by neglecting the fundamental, often unpredictable, discoveries from which future technological paradigms are born. The danger, then, is that in harnessing science for immediate technological ends, we may inadvertently diminish its capacity for the kind of foundational breakthroughs that have historically reshaped both knowledge and society.

2.3.2 Impact on Society

It should also be remembered that the demand for scientific development in a particular area is also dictated by society itself through some kind of "technological need." This is not a skeptical thesis, which underlies a broader plan to undermine the foundations of the rationality of science (and which for some readers may recall Feyerabend's thought); rather, it is an acknowledgment of how things sometimes go.

An ironic but fitting example was told to us by physicist Samuel Braunstein, now a professor at the University of York. The Berlin Wall fell in November 1989. Until those days, many people in the defense sector essentially owed their jobs to the Wall and the Cold War. Something similar was happening in the American university environment related to various fundamental sciences and engineering. In 1989, things, or at least their perception, changed drastically and suddenly. Panic nearly set in at the thought that, all of a sudden, certain fields of research considered important would be abandoned overnight: no more declared enemy meant no more reasons to develop new military technology and, consequently, no more reasons to push scientific progress. Within a couple of months, it became complicated to find an academic position in the United States in areas related to physics, a field that was somewhat in vogue, mainly due to the chaos generated by the perception of geopolitical change. Even job announcements for a position "in the middle of nowhere," at minor universities located in unattractive towns, began to generate fierce competition among a crowd of applicants. Braunstein himself, for example, applied for a temporary research position in a semi-unknown place; a position that, under normal conditions, would have been seen as unattractive. After a few weeks of silence from the potential employers, Braunstein tried calling. They told him they had received hundreds of applications and didn't quite know what to do. They were in an unprecedented situation because when new positions opened, they were used to receiving only a handful of applications, usually internal ones.

The impact of the complex dynamics associated with societal behavior is perhaps even more notable when we turn to technology. The sprawling influences of financial markets on the success of a company, and vice versa, the choice to develop or diversify a specific product or service, investor expectations, the stories told around a product and its use by customers (or consumers), the positive or negative perception of people, ethical discussions—these are all aspects that today are impossible to separate from what technology is. Certainly, one can argue about abstract aspects of technology, which may include definitions and reasoning, about its deep nature, independent of specific cultural contexts.

Soon enough, even starting from afar, we come to observe how the history of technology is dotted with irrational or difficult to understand developments, which must be seen in the context of what is related to society and, more generally, the human condition. We will return to this in Chapter 4 when we try to characterize the most important aspects of using technologies for the production of scientific knowledge and thus try to improve this process. In doing so, we will also discuss forms of *technological drift* aimed at maximizing aspects that do not appear "sustainable"—according to an understanding of the expression that we will adopt.

In the end, what this exploration reveals is that for all its logic of functions, formalization, and engineering, technology remains captive to the human condition. We have seen its path shaped by serendipity, its success determined by narrative, and its direction dictated by the often-irrational tides of markets and geopolitics. Technology, therefore, is more than a rational enterprise of problem-solving; it is a mirror. In the machines and procedures we build, we ultimately find a reflection of our own collective desires, fears, and ambitions, making the study of technology an intimate study of ourselves.

The complex relationship between science and technology is thus more than a historical curiosity; it represents a fundamental shift in human agency. Science, at its core, is a quest to *know* the world as it is. Technology is the act of *remaking* the world into what it could be. This chapter has charted the complexities of that act. The deepest reflection, therefore, is on the immense responsibility that comes with the power to make. Having acquired some ability to harness natural phenomena, the enduring challenge is to harness our own intentions, ensuring that what we build does not outpace our wisdom to manage it.

CHAPTER 3

Making Science

Notes on How to Support Research while Avoiding Bias and Gossip

In the previous chapters, we highlighted the most salient characteristics of science and technology for discussing the transformations they are currently undergoing. As we have mentioned in various ways, these are ancient, diverse, and sometimes conflicting activities, the result of a complex cultural evolution. If we have written this essay, it is because today, many people are calling for the urgent renewal of the scientific enterprise. Thus, we move from words and definitions to actions, addressing the critical points of doing science today and asking what form this activity might take for future generations. All in all, the adage that we must care for a world we will never see (often attributed to Bertrand Russell, though not confirmed by other sources) may sound sentimental, but it is worth embracing.

If we were asked, "What do you enjoy about doing science?" we would answer, without hesitation, that what motivates us daily is the urgency of collaborating with others toward a goal that, just like art, gives meaning and purpose to human experience, to echo a trope of Victor Weisskopf (1988). Of course, there are exceptions. There is a mythological conception of solitary science, exemplifying the traditional idea of genius: Galileo, Darwin, and even Einstein. Today, however, a different conception is taking hold, where the vast majority of scientific research revolves around groups that gradually come to dominate a field (see Wang and Barabási's book [2021], which we will return to later in this chapter).

DOI: 10.1201/9781003676614-4

The aspect to emphasize here is the importance of working with others in research, and the necessity of relying on the work of others. Many hands make light work, as the saying goes. It is all too easy to deceive oneself when thinking alone. It's easy to overlook details, to miss crucial aspects, and even to believe that something important has been accomplished when, in fact, we have barely scratched the surface of the problem. Dialogue, meetings, and exchanges of ideas are essential. This is yet another reason why science requires language.

Without delving too deeply, two main, and evident, challenges define the practice of science today: scientific publications and research funding. We will discuss them in order, considering alternative models.

3.1 PUBLISHING TODAY

3.1.1 The Primacy of the Scientific Article

Science is a craft, one that has evolved radically over the centuries. Yet, it remains a trade learned through a long apprenticeship within a group, absorbing explicit and implicit norms; cutting one's teeth, so to speak, in pursuit of what has become, in recent decades, an indispensable goal for maintaining or advancing one's position: publishing scientific articles in leading journals.

It is impossible to overstate the importance of the literary form of the scientific article when discussing the history of contemporary science and how science is conducted. Today, scientific articles can be seen as synonymous with science itself and are, in fact, its most fitting synecdoche. It is worth quoting Ginestra Bianconi's words on the importance of this literary form:

> In my opinion, scientific publication is a paradigm that has historically worked across the centuries, and I wouldn't change it. The body of research must become part of the world's cultural history. The access to online journals that we've seen in the last twenty years helps in this regard. Science is now considered more public domain than it was when there were only libraries. It's always important to have the ability to find an article, even one not heavily cited, years later, and perhaps discover a line of research that has been overlooked until now. The 'sleeping beauties' are research lines that, for some reason, did not develop immediately after publication but have the opportunity to flourish after many years. Other paradigms for making scientific results public may not have this advantage because an article can

contain nuances that even the author may not notice but can inspire those who read it with the intent to develop new science. Finally, the act of writing the article itself helps to expand an idea. Sometimes, a simple intuition requires extensive writing. Sometimes, you don't realize which aspects of the idea need development until you write a coherent article from start to finish. Writing an article in its entirety, therefore, helps to develop the idea itself.

Yet, the article is a recent literary genre, much more so than one might think. After all, the very image of science and the relevance of scientific communication have not always been as central as they are today.

Until the end of the seventeenth century, approximately, there were no scientific journals. The primary means of disseminating ideas were letters: think of the famous correspondence of Descartes, Galileo, or Gassendi. To a lesser extent, books were written. The first journals appeared in 1665 with the publication of the first volume of the *Journal des sçavans* (later *Journal des savants*) and the *Philosophical Transactions of the Royal Society* (Simone, as a curious fact, was a member of its editorial board). In her 2015 book *Making Nature*, Melinda Baldwin traces the history of the journal *Nature*, documenting how, as late as 1924, the term "scientist" in England had a partially derogatory connotation, paralleling that of "artist," to the point where the renowned physicist Norman Campbell had to write to *Nature* requesting to be considered not a "man of science" or a "scientific worker" but a "scientist" (2015: 4).

Now, following the detailed history of the scientific article would lead us too far afield. Let's focus on today and, for example, put ourselves in the shoes of Gaia, a young scientist who has just started her research journey (e.g., a three/five-year PhD or a two-year postdoc). Today, Gaia's research is measured by the scientific articles she manages to publish: this is the primary, if not almost exclusive, vehicle for disseminating the results of her work. It is within an article that we find, or should find, indications of the value of Gaia's research, the giants on whose shoulders it stands, and the new paths her results can open.

This complete reliance on a single literary form raises some questions. What, precisely, is this document that holds such sway over a scientist's career and the dissemination of knowledge? How has its structure been standardized to ensure the reliable transmission of information across a global, and often anonymous, community? To understand the pressures and incentives shaping Gaia's career, we must first deconstruct the scientific article itself—its form, its content, and the complex social machinery that brings it into existence.

3.1.2 The Literary Form of the Scientific Article

What is a scientific article? And how does publishing an article make a difference in the career of our hypothetical scientist, Gaia? The literary form of the scientific article has been structured over time to ensure the reliability of the message to the reader. What matters in this form is not only the contents, but also the rhetorical form in which they are conveyed, who conveys them (i.e., the prestige of the article's author), and the reputation of the journal in which the article is published.

A scientific article, the so-called paper, follows a fairly standardized model. It must have a title; it must list one or more authors; it must indicate the institution on whose behalf each author conducted the research presented in the article; it must contain an abstract, i.e., a brief summary of the article's contribution to scientific research; it must contain a text, typically structured into a section introducing the research followed by a presentation of the methods used and the results obtained; it must include a bibliography.

Each discipline, sub-discipline, or research field then adds other characteristics or presents some deviations from this standard. For example, appendices containing proofs or data may be attached, following established methods in the relevant field; or there may be photos, graphs and diagrams, or certain empirical evidence considered methodologically crucial to support the conclusions.

In short, a scientific article is a text written for an expert and demanding audience. It is essentially a working document aimed at reliably disseminating important information for advancing scientific knowledge within a community of people who may not necessarily know each other.

To advance her career as a scientist, Gaia will propose herself as the author of an article, but it is highly unlikely that she will be the sole author. Most articles are co-authored, resulting from the collaboration of a group whose size can vary enormously: from two or three people to thousands, as is often the case, for example, with articles on high-energy physics published by CERN (the European Organization for Nuclear Research). At the time we are writing, the highest number of signatures on a single scientific article is 15,025 (COVIDSurg Collaborative and GlobalSurg Collaborative 2021).

Gaia's first step, therefore, will be to establish her position as an author. If it is research that she has led, then she will be the first author; if Gaia has provided some data for research already started by a colleague or "contributed secondarily" in a vague way, then her position might be as a second or third author, etc.

If Gaia were the principal investigator (PI), that is, the lead researcher of the project, in charge of supervising and coordinating research activities, then she would be the last author on the list. The PI is often also the researcher who applied for and secured funding for the research on which the article is based. Thus, in some scientific fields, being at the bottom of the list is a source of pride and demonstrates a high social position. Since Gaia has just begun her research journey, she will most likely not be at the bottom; but, she can aim to be the first.

The order of authors in an article is not always peaceful and often leads to negotiations and, in the worst cases, real disputes, to the point that even *Nature* published some advice on how to manage authorship conflicts (Fleming 2021).

It all begins in an office, at a conference, or on one of the many online communication channels. Uncertain ideas gradually take the form of structured conjectures and theories through practices that are continually evolving, consolidated over time, and often specific to a discipline or field of study.

Once the theme has been identified and the people involved in the writing process have been defined, the next step is to choose what type of article to write. Simplifying a more nuanced landscape, there are three main types of scientific articles: *review articles*, which provide a summary of sub-areas of science, often accompanied by personal analysis; *research articles*, published in specialized journals; and articles published as *conference proceedings*. Each of these three categories has its own specific *raison d'être*. There are also other categories, such as opinion articles on already published results, but these are usually rarer and written almost exclusively by invitation. There is also the so-called addendum, where errors in an article written by the same authors, usually in the recent past, are corrected.

Review articles are useful for introducing a certain field of research or summarizing the state of the art in a particular discipline; they help those approaching that field or assist the scientific community in understanding the most important problems to consider, thus drawing researchers' attention to them. For example, the journal *Reviews of Modern Physics* publishes only review articles written in a style accessible even to non-specialists.

Articles of the second type, those appearing in scientific journals, typically require some time to pass from the moment the writing is completed to the moment the article is finally published in the journal. A 2021 study, limited to the biomedical field, shows a time interval ranging from 91 to 639 days (Andersen et al. 2021). This interval's duration, in any case, varies depending on the journal and the research area and largely depends on how long it takes for the article to pass through the so-called peer-review process, the evaluation process by experts, which we will return to shortly. For this reason, today, a significant portion of an article is disseminated through the so-called "pre-publication": the article is inserted into databases accessible via the web, such as arXiv or medRxiv, allowing the scientific community to read it and potentially use it for their research before its actual publication in a journal.

Putting scientific articles in a database offers significant advantages. First, it substantially reduces the time between when the article is written and when it can be read. Additionally, many databases (like arXiv) offer free services for both those who submit articles and those who want to read them. The value derived from this is obvious: no one has to spend a dime to make the scientific article public or to read and benefit from it.

The third category includes articles published in conference proceedings. The dissemination of these articles is faster: conference requirements are often less rigorous and meticulous than those of a peer-review process; however, even in this case, selectivity depends on the research field and the conference in question. In theoretical computer science, for example, the Annual ACM Symposium on Theory of Computing (STOC) and the IEEE Annual Symposium on Foundations of Computer Science (FOCS) are considered the most important events. The peer-review process, in both cases, is generally meticulous. Publishing an article in the proceedings of STOC or FOCS is usually more challenging and prestigious than getting it accepted by a journal. After the conference, an extended version of the article may also be published in a journal over a longer period.

In conclusion, when writing her scientific articles, Gaia will select the type that best highlights and values the lines of research she has pursued. This is because the article's function is to disseminate a result, sharing it with an audience that can use it for further research. This audience consists of present or future people who may be anywhere. Disseminating, in this context, means making the results of one's research known to other members of the scientific community, who will benefit from using them as part of their research. The greater the benefit, the higher (at least in some respects) the value of the research. This brings us back to the idea of a result's impact: those who benefit from a result will have access to new information that can be used to expand knowledge. This information can consist not only of problem-solving but also of introducing new techniques. Indeed, some journals provide referees (the next topic on the agenda) with a questionnaire that includes questions such as: "Does the article solve a known problem or propose a new technique?"

However, it is important to note that dissemination is different from popularization: scientific popularization is a body of work, often written in the form of an article or book, or conveyed through a podcast or documentary, aimed at inviting the general public to appreciate certain areas of science and, why not, to be inspired to follow its developments more closely. We will return to this topic in Chapter 4.

It is interesting to observe how the value of an article is connected to the impact it has on the community of researchers in its field and other interested communities. To determine an article's value (immediately and often even many years after its publication), there is no external criterion applicable to the dynamics of these communities. Scholarly impact depends crucially on the opinions of community members, which will form based on how the article in question affects their work: can these scientists continue to study the same thing as before, or, in light of the results proposed by the article, must they find a new problem? Can they now use a new tool to solve a problem they have already scrutinized and, in doing so, finally move on to another topic? Almost as if a secret has been revealed: the revelation leads to something "actionable"—an action that can be taken based on the information gained from the secret. Looking at this idea from another angle, the information gained reduces the degree of uncertainty encountered in answering other scientific questions.

There is no need to say that absolute transparency and honesty are required from those producing scientific publications. And it is interesting to note how dishonesty, often driven by behaviors accepted by the community, can manifest on many different levels, far beyond the mere fabrication or falsification of results—the most blatant expression of dishonesty. A less "serious" but widespread case is inflating the number of articles with the so-called *salami slicing method*: a scientific result is divided in such a way as to be published in multiple articles, when one would suffice.

Writing an article, however, is only the first half of the battle. For Gaia to have her work officially recognized and enter the permanent scientific record, she must now subject it to the scrutiny of her peers. This formal process of evaluation serves as the primary mechanism of modern science, designed to ensure quality, originality, and correctness before publication.

It is this ubiquitous, and often fraught, journey of peer review that we turn to next.

3.1.3 The Peer-Review Process

Now, as we have mentioned several times already, in order to publish her article, Gaia will have to go through a so-called peer-review process, or refereeing. The significance of the peer-review process can hardly be overstated. Its configuration varies considerably across journals, disciplines, and submission types (e.g., review articles, original research reports, or brief commentaries). In what follows, we sketch a generic version of the process. This overview is intended to demonstrate the intricate operations of the contemporary publishing system as well as the structural vulnerabilities to which it remains susceptible.

Once the article is written, our researcher will choose the appropriate journal to which to submit it. The article will usually be submitted to a journal by filling out a form on the web page of the organization that manages and publishes it. This organization can be an institution, but also a full-fledged company, with obvious financial aims.

In the next step, a junior editor of the journal will present the article to a person who acts as an editor. This person will be an expert in the field, able to determine whether the quality of the scientific article meets the level of the journal. If no editor can claim to be able to judge an article, even superficially, then this could be a sign that the article is not suitable for the journal. But, of course, there can be exceptions.

Once the article has been examined, the editor will decide whether to reject it or whether it deserves a more accurate, in-depth examination. In the latter case, the editor will take care of identifying some referees to handle the review. This is a difficult and delicate task, which in itself would be enough to explain the importance of editors for the world of science.

Generally, each article is evaluated by at least two referees, who have the task of reading it and, within a set time (usually between one and three months), sending a report that indicates the quality of the article and assesses whether it deserves to be published in the journal. The referees must be experts on the topic, but they must also not have a conflict of interest with our researcher. Here, the circle of people who can perform the task narrows. Not to mention that some people, perfect on paper to act as referees, may not be available (for example, because they are already too burdened with work, or because they are carrying out other institutional or political assignments). Usually, and this is no secret, if the journal allows them to suggest referees, then the authors will try to choose friendly people who are familiar with their work and, perhaps (if this is allowed by the journal), have co-signed other articles with them in the past.

The most widespread evaluation process is double-blind: those who evaluate do not know who wrote the article under examination; those who receive the evaluations do not know who formulated them. This also means that, in a double-blind process, the task of evaluating an article should not be accepted by a person who knows the author's identity. Obviously, in case the article is published, those who evaluated will become aware of the authors' identity; however, they will never be informed about who did the evaluation (unless this person decides to come out in the open). In other cases, instead, the process is single-blind, that is, it is only blind for the authors: those who evaluate them know their identity. Many research funds, such as those managed by the European Research Council, adhere to this rule, for example. In other, rarer cases, not even the editor knows the identity of who sent the article: it is therefore a triple-blind process, where both the editors and the referees do not know the names of the authors, and the latter do not know who the referees and editors are.

The choice of which degree of blindness to adopt is complex and depends on the purpose of the refereeing and its social context. The blindness of the evaluation process of an article aims to mitigate prejudices toward those who wrote it, both during the evaluation (double-blind process) and during the choice of who should evaluate it and on the final editorial decision to publish or reject it (triple-blind process); at the same time, blindness also protects the possibility for those who evaluate and the editors to exercise their judgment without fear of repercussions. Obviously, whatever the degree of blindness, we can imagine that the system ultimately turns out to be flawed, for example, because it is partial toward a certain social group or certain types of articles, or because it is random.

In their report, the referees will typically evaluate the innovation introduced by the article, the correctness of the results or the reliability of the data, the validity and originality of the techniques and methods developed to obtain the results, and the general quality of the manuscript, such as a correct list of citations. Our researcher then receives the (anonymous) reports of the referees along with their recommendations to the editor on whether or not to publish the article in question.

The recommendations tend to divide the articles into four categories. (1) "Ready for publication": obviously, fantastic news for the authors, who will see their article published as soon as possible. (2) "Containing minor imperfections": in this case, it is assumed that, with little work, the article can be adjusted and therefore published. (3) "Containing substantial imperfections": in this case, the authors will still have a lot of work to do if they hope to be able to publish the article. (4) "Not publishable."

An article cannot be considered accepted until it is declared category (1), i.e., publishable. Even a category (2) article could, in theory, be subsequently judged unpublishable by the journal.

In case of a negative verdict, one can appeal to the editor by writing a petition letter. Moreover, it is a good (and often required) practice to send a response letter to the referees every time a new version of the article is submitted to the journal: the letter will explain, point by point, how the authors have taken into account the referees' opinions in revising the article and, if they have not taken them into account, motivating their choice. The referees, for their part, will take this letter into account in evaluating the new version of the article.

Once the article is accepted, the text will be, so to speak, crystallized for dissemination, that is, frozen forever in the form it will have assumed at that point, and it will be perfected in all editorial details before setting it sail on the high seas of scientific discussion. The text will therefore have to be adapted to the editorial lines of the journal and de-anonymized, re-inserting the names and affiliations of the authors, acknowledgments to collaborators, and recognition of grants and funding received in support of the research. Finally, the authors will receive the proofs of the article from the editors who deal with this, and once the proofs are revised and approved, the article will be officially published.

If we have bored the reader with all these details, it was to give an extensive idea of how complex the publication process can be, an idea of all the infinite small negotiations that make up Gaia's work. Besides the fact that, obviously, we have omitted many possible complications, it should also be kept in mind that the refereeing process, although standardized, is in constant evolution. *It became common practice in the 1970s.* In the last two decades, it has been completely digitized with the spread of online journals, which make articles available with a click and allow text searches in a few seconds. Even in its digital form, however, refereeing is a laborious and complex process to manage. This is evidenced by the emergence of digital databases (the main existing database for science, arXiv, dates back to 1991), which, to make a certain result publicly available, do not require any type of monitoring or guarantee mark from the scientific community.

3.1.4 The Value of Publication

With a new article accepted, Gaia can now update her CV and inform her colleagues of the good news. Another small, important step in her career. What an effort! It's worth celebrating… what, exactly?

Publishing articles very quickly becomes the very life of those who do science. Every young researcher, like Gaia, needs to publish some articles, in a number that varies depending on the discipline and type of research, in order to obtain a PhD or guarantee the continuation of their career.

In fact, one spends a lot of time, when doing science, writing publishable articles. Paradoxically, one can also spend more time writing scientific articles than working on the research that supports those articles. Again, this depends a lot on the field, but it's not an absurd statement. The machinery behind the publication of articles is gigantic.

First of all, the scientific publishing industry employs thousands of people and constitutes a monetary turnover, on a global scale, of billions of US dollars. This fact is seen as problematic by some; see, for example, the protests raised by Marti Hagve (2020) on *Tidsskriftet*, the journal of the Norwegian Association of Medical Studies.

After all, there is a big difference between publishing a scientific article on a digital database and having a scientific article printed in a journal. The journal gives a quality guarantee, and, in some cases, the quality is certified by a real brand managed by experts in a given sector. Each research group has in mind a hierarchy of scientific journals based on the prestige associated with each; at the same time, every committee that hires new research forces will notice in which journals the articles of those who apply have been published. Some journals like *Nature*, *Science*, or *Cell*, then, are central to the careers of those who do research in any field.

In an attempt to bring order and standardize the value that is assigned to a journal or a scientific article by a community in the world of research, various parameters have been introduced. As far as scientific articles are concerned, the most important parameter is the number of citations received following publication; this number, obviously, also depends on how many people write about a certain topic. Over the years, other parameters have also been defined to generate rankings of scientific journals, both in general and for specific areas. One of the most used is the so-called impact factor. This is an index that quantifies the average number of citations obtained in a certain year by articles that were published in a given scientific journal in the two years preceding the measurement. Another parameter often used is a classification of journals issued by individual governmental bodies, such as the classification of the Norwegian Academy of Sciences. To evaluate scientific articles, then, one can rely on parameters such as the quality of the journal (through its impact factor or its classification by a governmental body), as well as the number of citations that the article receives.

It is, evidently, a jungle of numbers, in constant fluctuation and negotiation. Since the career of those who do research hangs on one of the myriad threads of this jungle, it's easy to understand the pressure exerted in recent years by these evaluation tools. If we add that the research contract does not provide for a minimum or maximum number of hours, and that many researchers can carry out their work 24 hours a day every day of the year, then it's understandable why doing research has become a very intense job, even from a psychological point of view.

Is it worth it?

3.2 LOOKING FOR ALTERNATIVES

3.2.1 Against Peer Review

Peer review has come under severe criticism. It is important to consider them at this point. Bias is perhaps the most obvious and, to some extent, unavoidable of the issues. Referees may tend to evaluate papers from well-known scientists more favorably or may be more accommodating to articles that align with their own viewpoints, even down to stylistic preferences that reflect cultural, gender, or linguistic traits. Biases and prejudices are human phenomena, natural expressions of our psychology, and they are difficult to identify and control. Indeed, the literature on bias in peer review is divided on its pervasiveness: in a highly cited and controversial 2012 study, Carole Lee and colleagues (2013) suggested that there may be less bias than commonly believed. Yet, in subsequent years, numerous articles have documented a long history of gender and linguistic biases (see, for instance, Llorens et al. 2021; Politzer-Ahles et al. 2020).

Can we define methods to reduce biases? Double- or triple-blind peer review is one attempt to do so. Yet, the entanglement between bias and expertise can hardly be completely undone. Not to mention the fact that experts in a certain field often find it easy to guess who authored a particular paper.

One way to address the issue, already adopted by some journals, is to experiment with a different type of peer review, known as "open peer review." This approach is based on the transparency of the process, making the names of reviewers, their comments, and, when possible, the discussion between reviewers and authors public. Open peer review has its advantages: it attempts to avoid reviewer partiality, allows for further interactions within the scientific community, encourages authors to think twice before submitting low-quality work, and reduces nepotism and favoritism. However, it also has drawbacks. First, the academic hierarchy tends to discourage junior members of the community from publicly criticizing senior colleagues, thereby limiting the potential for open and fearless discussion. Additionally, it's important to note that such open discussion would likely take place through digital means, which could introduce impersonal elements that alter the communication process in various ways—for example, written text can easily lead to misunderstandings. Finally, a reliable strategy would be needed to build and maintain trust in this new system. In short, while open peer-review is an interesting proposal, it certainly presents some challenges to be implemented.

Adding to these critiques and difficulties with peer review is the "cherry on top": many articles, once accepted, can only be published if the authors pay a fee to the publisher. Moreover, the published article may not be freely accessible to any reader unless an additional fee is paid to guarantee open access through the journal's website and other channels. This mechanism exacerbates disparities in research on an international level, favoring the strengthening of certain institutions (and their employees) at the expense of others.

The disparities can then quickly scale up for a researcher, a research team, or a research institution. The field of bibliometrics attempts to quantify the value of scientific research, such as the value of individual articles or journals. However, in its attempt, it can pave the road to create paradoxical situations where science is no longer pursued to obtain meaningful results but rather to produce publications: of essence is to bypass the peer-review process. The phrase "publish or perish" sarcastically illustrates this reality, and it's saying that this phrase is also the name of a software program designed to analyze citations (Harzing 2007).

After reviewing these problems associated with peer review, it becomes evident why many in the scientific community have questioned the continued use of this system. Tradition is not a sufficient justification: while peer review has become an arbiter for competitions, career advancements, salaries, and reputations, its flaws and inherent biases are undeniable, not to mention its economic, intellectual, and social inequities.

This widespread dissatisfaction with the status quo has naturally fueled a search for alternatives. If the current system of gatekeeping is so flawed, perhaps the solution lies not in simply tweaking the review process, but in fundamentally reimagining the very artifact being reviewed. This leads to a radical question: what if we were to change the literary genre of the scientific article itself, moving toward more dynamic and collaborative models of communication? One such proposal, the "open-ended" scientific article, offers a potential path forward.

3.2.2 Open-Ended Scientific Articles

One strategy to address the problems of peer review is to change the literary genre that characterizes contemporary scientific research and, more generally, to reinvent the way science is communicated. Sergei Kalinin, Weston Fulton Chair Professor at the University of Tennessee in Knoxville, has described a proposal in this direction. We'll summarize a dense conversation we had with him about so-called "open-ended articles," an idea Sergei has supported along with his colleague Andrew Akbashev in various writings (for a starting point, see 2023). The good ideas here are Sergei's; any mistakes are our own.

Sergei explained that one of the problems with science today is the separation between the measures of success in scientific research and the tangible results that lead to technology. These results, after all, are one of the main reasons society, through funding agencies, supports the scientific enterprise: there is a desire to have some impact on the well-being of citizens, perhaps in specific regions of the world, or to contribute to energy needs, health, culture, and so on.

According to Sergei, science should not exist as an activity for its own sake. Yet, it is a challenge to strike a balance. The question to ask is: what is the measure of success in science? Science always tries to solve problems related to the real world, but the timeline for the impact of solving these problems may be distant. In our attempt to understand how the world works, basic science, in the long term, also helps solve short-term problems, but this happens where the "short-term" window can be placed in the future, even in the distant future. The timeline for what we call applied science, on the other hand, is closer.

Focusing on the task of evaluating scientific production, we are faced with a main problem: publications are a quantitative performance indicator in the short-term context. The risk is that the true impact of science is replaced with an essentially promotional element.

Sergei's idea is that articles should evolve: the standard model should be of an "open-ended" body of work. A scientific article should, according to Sergei's proposal, be understood as a work in progress and, therefore, open. We can draw a parallel with recipes. A recipe—for instance, for trofie al pesto—is not a finished idea once and for all: as utensils and the spaces in which we cook evolve, as ingredients, needs, and conviviality change, trofie al pesto are adapted and reshaped. In essence, we have an evolutionary adaptation of a recipe that follows, also, an adaptation of our knowledge and needs (see Borghini 2015). Similarly, a scientific article could be considered an artifact that evolves analogously to a recipe.

A proposal going in the direction suggested by Sergei is the *Lifecycle Journal*, an initiative of the Center for Open Science, a non-profit organization based in Washington, DC. The *Lifecycle Journal* adopts a "publish-before-review" model:

> the review process consists of several independent traditional human peer review, machine-based, and empirical credibility assessment services. *Lifecycle Journal* and the evaluation services do not issue accept or reject decisions. Instead, authors are empowered to decide when they are finished with the active phase of revising and responding to evaluations by optionally declaring the present version as the Version of Record (VOR). The VOR marks the end of the journal-coordinated review process and is akin to a traditionally published journal article.
>
> *(see https://lifecyclejournal.org/)*

At this point, however, we have a problem: the open-ended article model just described becomes a metaphor for science itself and, at the same time, a never-ending project, like the construction of St. Peter's Basilica. But this shouldn't scare us: we must think of science just like recipes, even abandoning rigid models of authorship—a theme on which thinkers like Foucault and Barthes have already reflected some time ago. We can list the people who are the original authors; then many others can collaborate to improve, review, update, and adapt the work. Just as we do with trofie al pesto. In the case of science, taking care of an article will always aim to improve its epistemic profile: we want to maintain or increase its reliability, updating it based on new discoveries and connecting it to them.

Scientific articles, of any type, exist because scientists need to share their ideas and discoveries; they appeared at a time in history when the only way to do this was to use written documents as a method to archive and discuss results. And this was, in fact, a very effective way to achieve the goal of sharing. Today, however, different and additional needs have become important. We need ways to connect individuals who obtain results within a much larger scientific community (and, also because of this, communication and information-sharing options have grown in the last 30 years). Additionally, there is much more to scientific research than just publishing results: for example, consider the importance of databases that make the data on which research is based available to the community.

First of all, sharing data is not enough to guarantee knowledge growth. Qualified experts are needed to associate meaning with the data. Sharing the raw data from my experiment would likely be useless, except for the ten people in the world who are doing similar experiments. It's important to share the entire workflow, that is, the entire process that leads to the production and processing of that data. Making this process transparent builds bridges that connect narrow domains to the broader scientific context; it also fosters machine learning in ways that have yet to be invented and determined. To conduct research, whether in physics or the natural sciences in general, it is necessary to share the entire process that leads to the production of data, including a machine learning model capable of imitating that process and, therefore, of formulating ideas, suggestions, and even selecting the experiment (what results do we want an experiment to lead to? Why do this experiment instead of another?). A separate discussion should be held for mathematics because a result is not determined to be right or wrong based on an experiment, but solely on logical constructs. Data science, on the other hand, begins where the experiment ends.

Second, to conduct research, it is crucial to formulate a hypothesis because one cannot simply conduct an infinite number of experiments on everything. Instead, it is necessary to determine the greatest source of uncertainty in the "worldview" we intend to build and go directly to that point to conduct the experiment. The experiment can be conducted by a machine or a human agent; it doesn't matter. In either case, there will be two possible outcomes: the first is the falsification of a hypothesis; the other is the beginning of a series of fortuitous and unexpected discoveries—the so-called serendipity we have already discussed.

Finally, it is important to note that, according to Sergei, machine learning alone will not be able to solve problems that require experiments. This is because if the experiment does not falsify a hypothesis, the machine will have difficulty determining the value of the information obtained in relation to the theoretical model, and it is inefficient to modify the theoretical model every time new information is acquired. To close this loop, machine learning and the experiment must converge with human intervention.

We conclude this account of the conversation with Sergei with the following (somehow unsurprising) suggestion (or hope): the future of scientific publications will involve sharing results, but also data, and especially the workflow that can accelerate the dynamics of this cycle.

What matters is that this proposal is not an isolated idea but part of a broader movement to reinvent scientific communication. This exploration includes many other forms, each challenging the static nature of the traditional paper. From *executable articles* that fuse narrative with live, reproducible code, to *nanopublications* that break research into atomic, machine-readable claims, the goal is to create a more dynamic and modular scientific record. Other innovations focus on the process itself, such as *registered reports*, which validate methodology before results are known to fight publication bias, and *community-driven wikis* that create a single, living source of truth for a field. Together, these alternatives point toward a future where scientific knowledge is less a library of finalized texts and more an interactive, verifiable, and constantly evolving network of ideas.

This vision of a dynamic, collaborative, and data-rich scientific ecosystem is compelling. However, by dismantling the traditional structures of publication and authorship, it forces us to confront a new set of challenges that are fundamentally social in nature. In a world without the clear signposts of journal prestige and finalized articles, how do scientists build a reputation? How is trust established and maintained within the community? And what prevents such an open system from being influenced by less formal, and perhaps less reliable, mechanisms like gossip and social networking?

3.2.3 Reputation, Gossip, and Trust

The problems of doing science are not only, so to speak, "internal": another very general problem is related to how science is viewed from the outside. To understand and predict the evolution of science, we certainly cannot only consider the truth of scientific results. Those who study the history and sociology of science have exposed the complex intertwining of the pursuit of truth and the context in which it takes place. In fact, science also depends on ideology (religious, political, theoretical, etc.), as well as on gossip and the reputation of people and results.

Today, the social context in which scientists operate is heavily influenced by certain technologies. The so-called "twitterization" of scientific research is indeed a phenomenon that places reputation ahead of results, exponentially amplifying the importance of the former over the latter. It no longer matters what is discovered, but that the news of the discovery spreads—no matter how sketchy, brief, and unverifiable it may remain. Announcements of discoveries made on social media or through brief press releases are increasingly common. The rest of the amplification game is played by the reputation of the journal, the institution, the authors, or the topic itself if it is particularly recognizable outside the scientific community. Attention to topics can have a short cycle. Sometimes the ball is picked up quickly, and thus even unclear and not fully verified results can be used as a promotional tool.

Ginestra Bianconi offered us a fine summary of this situation:

> What is my personal motivation? I do not seek to do science to find immediate success, such as financial success or self-promotion, but to seek personal satisfaction, expressed by being able to do what I find interesting. So, for me as a person, the priority is to pursue ideas and lines of research that interest me. On the other hand, however, there is a constant comparison with the community: in the way it interprets, understands, or even ignores your scientific work. This feedback proceeds through the publication of articles, and I believe it is an exercise that the community has always done and will remain valid in the future. The balance between your scientific interests and the feedback the community gives to your publications has some of the flaws of any society made up of human beings. It is subject to trends, for example. At this historical moment, a lot of energy is also spent on publicizing scientific results, and this sometimes takes precedence over the significance of the work: the quality of a scientific article is no longer evaluated on its own but also on how the article is perceived by the community. By doing so, even the desire for specific knowledge and individual motivation is influenced by the community.

But what are the topics that are most likely to capture attention? There is the attention of the scientific community and then that of the general public. The perception of the former tends to be less influenced by twitterization. There are still risks, such as steering science policies to support certain topics more, with consequences for funding allocation. More famous areas attract more funding even if they do not solve important or urgent social problems.

For the general public, a danger is that all this may influence the new generations in choosing research topics. We tend to develop interest in what we come into contact with most frequently. Topics that can be explained simply and evoke clear images have an advantage. For this reason, scientific dissemination plays a vital role.

3.2.4 Trust in Science as a Public Value

There is another side to the coin. The spread of news about scientific activity on the Internet risks translating the work of scientists into mere gossip, trivializing it, almost as predicted in 2010 by Nicholas Carr in his essay *The Shallows: What the Internet Is Doing to Our Brains.* How can we avoid this superficialization (and the often-correlated generation of distrust in science) among citizens and scientists themselves? Perhaps it would be better to let a machine select the direction in which scientific research should proceed.

Sergei Kalinin once again shared with us very interesting ideas, such as the problem of trust in authors. By definition, when we do science, we always move toward the unknown. Otherwise, it is not science. And when we move toward the unknown, we can expect to obtain results that will change the way we perceive the world. Science changes our worldview. If it did not change our worldview, there would be no need to do it. Once we realize this, interesting considerations follow, which we can try to rationalize.

The dynamic entanglement between scientific knowledge and worldview is fruitfully captured by the Bayesian approach. According to this, a theory is updated by an experiment in proportion to the probability of success of the experiment in question: the more likely it is that the experiment succeeds, the less of a change it will produce in our worldview, if the experiment succeeds. For example, if I can already expect that a glass will break once I shatter it, then shattering it will not alter my theories regarding glass. The very important thing about this approach is that it prescribes changing the worldview based on the degree of confidence in the result, based on past results. This is absolutely essential because it makes the process well-parameterized. Everything depends on the probability distribution before the experiment is conducted: if my prior distribution is very broad, I can believe everything—a condition certainly not ideal. If my prior distribution is exceptionally narrow, then I cannot believe anything: my opinion is fixed, immutable. Therefore, in general, the shape of the distribution represents my confidence in new discoveries: if I believe a lot in new discoveries, then I will change my opinion; if I believe very little in new discoveries, then I will not change my opinion at all—regardless of the results.

The problem is that, at least at this historical moment, the degree of confidence is a very subjective thing, even from a scientific point of view. Imagine reading a new result in a prestigious journal and finding it convincing. Then imagine that the same result was obtained by another research group, of which no one has ever heard, and which has no reputation in the field. It will be "natural" to believe in the results and opinions coming from the more established group and the more famous journal; but on what basis? We "naturally" place different levels of trust in results from different research groups. A significant discrepancy thus emerges between this image of science and the principle that science is independent of the social position (or status within the scientific community) of the person proposing it.

If we have a formal system of demonstrations, such as a mathematical theorem or a physical result that can be verified, then it is absolutely true that the person who has just started their scientific career or the person who has been engaged in science for decades has an equally unassailable result, regardless of who proved the theorem. However, if we consider factors such as the importance of a particular result, or whether this result is solid or not, then the group from which the result originates becomes extremely important.

The way we perceive new information and new discoveries depends on our trust and the source of this information. And, in a sense, we cannot build any kind of social organization, such as science, in the absence of trust. How far can we go without trusting the people who tell us they have made a scientific discovery? Not very far, obviously. But how far can we go if we completely trust everyone? The same situation. So, in a non-ideal world, building a system of trust is the only way to navigate and truly progress. Breaking the rules may provide short-term advantages. In the long term, however, creating trust is the real advantage. Here, an analogy with discussions about selfishness and altruism emerges: in the short term, being selfish—that is, caring only about our own utility—seems to be a better choice; but altruism can be the ideal choice when our time horizon is longer.

Promoting trust differs between short-term and long-term strategies. The violation of trust may be useful in the short term. For example, publishing a large number of mediocre but very visible articles when one is seeking a new academic position is useful. But, over time, research groups that build their reputation on trust will always prevail. So, all things considered, perhaps the way forward is to create incentives for long-term trust development while, to some extent, also giving priority to other short-term incentives.

We must also remember that, in general, in the scientific community, the incentive system must also be based on the fact that all human beings are essentially individual agents. For all of us, the balance of rewards is different: some people value the opportunity to do something interesting and enjoy freedom more than the financial side; others value the ability to do something that keeps them in the shadows, while others still like publicity.

In the end, rewards for different people are different, which is almost providential. How can we formalize scientific policies in such a way that people want to follow their own incentives? And so that they are rewarded when a certain condition of trust is met.

3.3 RESEARCH FUNDING

3.3.1 Support for Research, Quality, and Quantity

After a long excursus on peer review, we shall return to Gaia. Her career will crucially depend not only on the publication of scientific papers but also on the funding she can secure for her research. We have seen how complex and cumbersome the mechanisms governing the publication of scientific papers are, and we have suggested at least one way to revise them. What is the current state of research funding? How is scientific quality measured, and how can these measurements support the advancement of knowledge rather than self-promotion and scientific gossip?

We discussed this with Aram Harrow, a physics professor in the Center for Theoretical Physics at the Massachusetts Institute of Technology:

> We might measure the impact of science based on the technology it creates and how that technology is used. In other words, we could consider successful technologies and the science behind them. However, these matters are difficult to quantify. And if career incentives are at stake, then the metrics will be very individualistic. In short, sometimes the system makes mistakes. What I would like is a

> system capable of reviewing these mistakes, of discerning between what we understood well and what we got wrong, capable of making us understand why our funding system didn't allow us to see our mistakes earlier.

These observations confirm the impression that we are at a crossroads, at least concerning the topic we are discussing here: we are faced with the need to think more systematically and perhaps programmatically about how to promote the value of scientific research through funding. Aram continues:

> What worries me about science is that there isn't much introspection. Resource allocation, grants, and jobs are the main things to consider. There are signs that things are going wrong. Firstly, the so-called replication crisis in many areas of science. It's a methodological crisis. Too many experimental results are published but are not replicated or are not replicable. What does this mean? If there is a replication problem, it means that the people doing science are not truly interested in seeking the truth. They want—we want—to achieve something publishable and formidable, something sensational, rather than something reliable and useful. The culture we have created does not align with the goals of science. We give incentives to produce results that seem exciting but are potentially untrue and, more detrimentally, useless.

Adding to this, we can also cite the opinion of an equally authoritative person working in a field different from Harrow's, namely British biologist Paul Nurse (Koch 2013). According to Nurse, research funds should be focused more on people than on their ability to write grant proposals or on how much the content of those proposals aligns with what is "in vogue" at that particular moment. According to Nurse, interdisciplinary collaborations are more likely to lead to technological applications that benefit society. Moreover, the scientific community should listen more closely to the public to understand its concerns and how to address them.

The perspectives of Harrow and Nurse converge on a single, powerful point: the current system of funding often rewards the wrong things—sensationalism over substance, grant-writing skill over scientific potential. To correct this, however, we face a deeper, more fundamental challenge. Before we can fund better science, we must agree on how to recognize it. This forces us to confront one of the most elusive concepts in the scientific enterprise: the nature of a "breakthrough" itself.

3.3.2 Access Criteria: Degrees and Types of Breakthroughs

What has been said so far omits what is perhaps the main problem with access to funding: when we are dealing with a truly significant result, how do we recognize it? One term, more than any other, dominates the conversation on this topic: breakthrough.

Although the expression "scientific breakthrough" may sound familiar, it is not easy to define. We can characterize it in many different ways: "turning point," "discovery," "step forward," and "milestone." A breakthrough often connects the progress of science with technology and consists of an advancement of knowledge.

Today, we read daily in the media about important scientific discoveries. Unfortunately, though, it's essential to distinguish between those that are truly significant scientific discoveries and those that are merely attempts by academics, large companies, or startups to self-promote their research. A (rather eccentric) testimony to the necessity of this caution was the *quantum bullshit detector*, a Twitter account opened in 2019 to expose all the nonsense being written about supposed breakthroughs in quantum computing.

There have also been nearly embarrassing situations where it was evident that a non-discovery was mistaken for a breakthrough. A case that caused a stir in certain scientific communities dates back to November 2019, concerning a piece in *Quanta Magazine* (Wolchover 2019), which later led to the publication of an interesting review article in 2022 (Denton et al. 2022). A research group in physics had singled out an identity for calculating some quantities in linear algebra (eigenvalues from eigenvectors for a family of matrices). Believing it to be a true discovery and not a previously published strategy, the group began corresponding with several important mathematicians, including Fields Medalist Terence Tao, to test the waters. Tao responded that he had never heard of it, although the result had such generality and simplicity that it could already be present in matrix analysis textbooks (see Terence Tao's blog post "Eigenvectors from eigenvalues," dated August 13, 2019). Finding no references, the group concluded that it was an original discovery and decided to write an article about it.

What emerged in the weeks following publication is astonishing: cross-referencing bibliographic searches, it became evident that variations of the identity had already been published dozens of times; this so-called "discovery" even dated back to the work of German mathematician Carl Jacobi in the distant 1834! Clearly, even eminent mathematicians have difficulty retrieving widely published results.

In an ideal world, a primary difference between science and technology is that a scientific result should almost be able to self-evaluate. By this, we mean that once a result is obtained, it should be the scientific community that determines its value and assigns it a label. For example, it should be the scientific community that says a particular discovery is important—not the authors themselves, or the media. Examples like that of *Quanta Magazine* demonstrate the difficulty of systematic self-evaluation and of navigating the literature.

Let's broaden our perspective a bit to remind ourselves of a question central to much classical reflection in the philosophy of science: what characterizes an important scientific discovery? It's difficult to find a common denominator. There are scientific discoveries whose value is determined many years after the discovery itself. There are other discoveries whose value is determined immediately. Consider, for example, the proof of an important conjecture in mathematics.

A conjecture is a mathematical statement whose truth value is unknown. Once proven, it becomes a theorem. The transition from conjecture to theorem is fundamental. Not only because a theorem contains a proof, but also because it typically has consequences for other conjectures, which can themselves become theorems if that conjecture is proven. The value of a theorem can thus be determined by the impact of the knowledge whose truth value is determined by the theorem itself, knowledge for which we previously had no guarantee of truth.

There are also results whose value we cannot immediately determine. For example, the introduction of new techniques. A new technique is useful because it allows for further discoveries. Returning to the example of mathematics, it could be a new technique for tackling a certain type of proof or calculation that we were previously unable to do.

Consider a well-known problem that has long resisted the efforts of the wider mathematical community. Because a significant number of experts have grappled with this problem, its eventual solution is immediately recognized as a major advance. The value of the solution is thus determined by an almost sociological impact rather than a scientific one. In general, a discovery is important when it allows for other discoveries. When we ask a question about the world, the answer to the question has greater value when it allows for many other answers; in other words, when many other answers to many other questions are automatically determined by the answer given. The value is (also) determined by the technology that can be derived from the scientific discovery. The timeline and process of transitioning from scientific results to technological applications are difficult to describe. They follow dynamics that depend greatly on the historical context and the specific field of research. Consider, for example, biomedical sciences, artificial intelligence, and aeronautics: very different fields, with innovation cycles of varying lengths and particular characteristics.

But, returning to the theme of this section, who judges whether a discovery is a breakthrough? If we wait a certain amount of time, it should become evident. Today, through the media, we see situations where attempts are made to distort the perception of the value of a scientific discovery by telling a story that does not fully reflect it—a parallel story, told for reasons of scientific marketing or self-promotion. This is done to gain attention and create the perception that a scientific discovery is important without waiting for the time it takes for the scientific community and society itself to absorb it and use it both in science and technology. The narrative used alongside the scientific discovery conveys a potentially dangerous message because it mixes reality with observations of a fantastical nature.

Scientific marketing thus becomes a powerful tool, modifying reality and exerting control over how both the scientific community and society interpret science. The greatest risk is contributing to the creation of an imaginary reality in which science, while still an activity partly independent of human opinion, acquires subjective value. Another possible outcome of abusing marketing tools is increasing the notoriety of certain fields of research at the expense of others and pushing funding agencies to make decisions based on an imaginary reality rather than one connected to facts. Upholding factual accuracy in science thus becomes a moral necessity.

We close this section by noting that, with the growth of scientific communication and the importance of justifying the support received from often public entities, there is a strong temptation for every scientist to want to utilize various communication channels and share content that goes beyond the scientific article and their own results. However, truth and factuality have practical value that benefits science, in contrast to the harm that can be caused by distorting scientific results through imprecise, too brief, or self-promotional communications.

This pressure to generate and market breakthroughs doesn't just affect how science is communicated; it profoundly shapes what science gets done. When the incentives reward discoverable, easily packaged results, researchers may be subtly pushed toward problems that are more likely to yield such outcomes. This raises a critical question about the research environment itself: does the very context in which science is performed—the intense competition for funding and recognition—lead researchers to search for answers only where the light is brightest, regardless of where the most important questions may lie?

3.3.3 The Importance of Context and the Streetlight Effect

A problem perhaps more significant than others is the context that influences scientific research. We discussed this with Fotini Markopoulou-Kalamara, a person with experience in both physics and industrial design. The following is the result of our attempt to put order in the notes from the conversation. Again, the good ideas are Fotini's, ours are the mistakes.

Let's start with an apparently simple question: how do we determine whether a scientific result is correct? We need a validation mechanism that, however, is not always obtainable through mathematical tools. On the contrary, we often rely on the idea that the data contains the truth. Unfortunately, this is an idea more influenced by culture than by evidence. Data always has a context. When we examine data—in the form, for example, of a statistical observation or the reading of a scientific instrument—we must consider the scientific community, the source of the data (including the human and broader context that produced it), and the question we wanted to answer.

Often, how we formulate the scientific question is central. For example, consider the study of executive functions in cognitive sciences (Stuss - Benson 1986). These are the cognitive abilities a subject uses to regulate their behavior, such as volition, self-control, strategy planning, or gathering the necessary information to achieve a goal. It is common in this field to make statements like: "These are the data on situations where people pay attention to A, B, and C."

Therefore, we need to examine not only the data but also the context. Data of any kind has a context, a narrative, a story in which it is "immersed." The question to ask is: what does the story tell us? The construction of the story is a technology in itself. It could be called the technology of reality construction. In the above example of studying executive functions in cognitive sciences, for instance, it is a technology; an approach that, in the end, determines what people pay attention to.

We live in a cultural context very attentive to the visible, which, on the other hand, tends to disregard the invisible. It's often easy to convince ourselves that what we cannot see does not exist, and so, everything that is not included in the variables, in the degrees of freedom that we decide and can measure, almost disappears, becoming invisible. This also applies to the cultural context of the technology we live in today, in which we interpret data as if the invisible did not exist: since we cannot measure it, we behave as if it has no impact on us. Yet, some invisible things, like the variables that add "noise" (randomness and uncertainty) to a system, obviously exist. The entire context, in short, is not always visible. And it is not clear. It even becomes invisible when we take it for granted.

Every time we have data, there is a specific community that works with that data. For a given physical theory or a given approach in medicine, there will be a certain community for which that data makes sense. In the context of mental health, for example, there is a transition underway toward including patients (or at least the subjects whose data is then studied) in what is being measured or in what the study's goals are. It's complex to provide a single, systematic method to determine whether a patient has improved after a certain therapy. The answer to the question "Has there been an improvement?" is based on something measurable, but some new directions invite the patients themselves to choose the measures. People thus become part of the system, and the scientific challenge is to obtain results whose interpretation does not depend on those people. Unfortunately, it is unclear whether this is possible in general. In short, the scientific method itself tends to be contextual.

An additional problem—seemingly distant—is that much research is subject to the streetlight effect, as David Freedman called it in his book on failures in science (2010). This effect means that scientists tend to work on solvable problems rather than important problems to solve—like searching for a set of keys under a streetlight at night not because we lost them nearby, but because it's the only place with light. A question from which data can be gathered is not necessarily an important question. In studies on attention, for example, the streetlight effect is prominent.

It would be useful to define a parameter, let's call it the *L*-index, capable of determining how much the streetlight effect impacts a particular problem: the higher the index, the stronger the streetlight effect. We would then have the *L*-index of how much a scientific article tries to solve an important problem, instead of following a line of research where many people are finding incremental results. It is obviously a difficult index to define and, as far as we know, it has not yet been proposed or studied as of today (spring 2024). The difficulty lies in defining something that quantitatively captures the amount of innovation. It is an arduous task because, once again, it will depend heavily on the context. And if the mysterious *L*-index is too high, then the science will not even be understood by the community doing it.

Thus, Fotini continues, there is always a community that determines the context of "truth." Truth is not independent of the community to which it belongs. It must have a context. Knowledge cannot be associated with data but must be compared with people. It lives in people. In communities where tradition is valued, for example, elders are relied upon because they remember the reasons why we ask one question instead of another today. This is a form of technology for mapping reality. What we decide exists in the world, and what we pay attention to, depends on the technology we adopt, and this changes over time. Thus, what is visible or invisible becomes a cultural fact. It is impossible to remove these cultural aspects and prioritize science as an impersonal initiative, because one cannot work with data whose history is dissociated from people. The component associated with the community is crucial because it is the community that has reached an agreement, probably tacit, on the story to tell in connection with the data.

In conclusion, how does context influence discovery? Results can be falsified, but this is quite rare and difficult to do. However, one can try to pose a question as if it was the right question to ask; it's likely to be a question with a high *L*-index. There can also be situations similar to Columbus's egg, where a seemingly impossible problem turns out to be easy once understood, simply because no one had thought to look under a very bright streetlight. On the other hand, these situations are rare, considering the breadth of the sub-communities in today's science. It is more likely that, if we stumble upon a discovery, it will have a relatively low *L*-index because the context does not encourage moving away from the streetlight (indeed, it may even penalize those who do).

A very recent example is the experience of Katalin Karikó, a Nobel Prize winner in 2023 for her work on messenger RNA. In the mid-1990s, Karikó was penalized by her university because she was working on unconventional problems, to the point of being relegated to a laboratory on the outskirts of the campus and having difficulty publishing her results (Zuckerman 2023). Karikó's research was viewed with great skepticism by the scientific community because it was far from the mainstream. Fortunately (not only for her and the scientific community but also for society as a whole), Karikó had the strength and courage to keep going. It is clear, then, how a high *L*-index associated with particularly ambitious or unconventional research can be problematic.

3.3.4 How to Improve the System?

So far, we have highlighted several critical aspects of doing science today. We have not done so to lament but to think about how to correct the course, if possible, and to identify the obstacles. To briefly summarize, the activities of producing scientific knowledge are too often inefficient (wasting energy and ideas), erroneous (failing to eliminate incorrect results, unsupported by evidence, redundant, or even obtained with malice, whether conscious or unconscious), and unjust (failing to reward potentially most fruitful ideas). So, how do we repair and improve the system?

Metascience entrepreneurship is still in its infancy. It is not based, and perhaps should not be based, on a single method, much less on established practices. Instead, there are several interesting and credible proposals in the literature today whose results are, however, yet to come. We can distinguish four of them here, without claiming to be exhaustive.

The first method is the science of science, characterized by the use of network science or, more generally, the science of complex systems—a very broad field of research with branches in mathematics, biology, and the social sciences (Fortunato et al. 2018; Wang and Barabási 2021). This approach can offer a quantitative analysis of the historical trend of scientific research, which can be represented according to multiple parameters, providing a predictive basis that is undoubtedly innovative and easily communicable. Why? Because using metrics presents itself as a science with quantitative aspects that can be used to make decisions.

Then there is the evolutionary approach to scientific culture, which uses the tools developed in the theory of evolution (both natural and cultural) to understand and predict how science evolves (Haufe 2022). Currently, this approach is less precise and rigorous than the science of science because it is not as evident what the fundamental parameters are on which the cultural evolution of the scientific enterprise depends; presumably, they are deeper parameters than those used by the science of science, which quantifies based on surface variables such as the keywords associated with an article or the number of publications in a scientific area. In summary, there is a lack of data to follow the evolutionary approach fruitfully.

The third is an entrepreneurial approach that seeks to innovate the social relationships underlying scientific work and is particularly representative of Nielsen and Qiu's essay on metascience (2022). In this case, the solution to the most critical aspects of the scientific enterprise would lie in devising new principles to coordinate its social aspects and optimize its changes, for example, by pushing toward new approaches to funding distribution, rethinking peer-review processes and the settings of scientific journals, developing new models of collaboration for research in non-academic contexts, or those with one foot in the university and the other in the enterprise.

Finally, a philosophical-political approach to science reflects on the principles that should inspire those who govern it at different levels and those who regulate its various processes (a fundamental book in this field was Philip Kitcher's *Science, Truth, and Democracy* [2001]).

Based on the insights from these four very different methodologies, as well as adding a few of our own ideas seasoned with thoughts gathered from various interviews, we begin to imagine an alternative model of doing science. It is, of course, a utopia, an exercise to try to imagine a better situation, which will occupy us in this and the next chapter.

Let's be guided, for the discussion, by an idealized scenario. Suppose we establish a metric to determine which of two or more competing research projects deserves to be funded based on how much the project promises to benefit society. One of the problems associated with this goal is defining, with some degree of precision, what is beneficial to society. Which society are we talking about? Should we look at the short term or the long term? Are there groups of people to whom we should pay more attention? When is it justifiable to take a partial stance? And, above all, is it possible to answer such a generic question, given the risk of falling into the vacuity of futurology? For example, if the funding choice is made by an industrial entity, whose interest will be at least partly linked to profit (even decades from now), then the research will end up benefiting specific groups (for example, customers with access to more monetary resources and already close to the specific technology) and discriminating against others; can we accept this, and if so, to what extent?

These are well-known questions for those involved in metascience, and we have no magical solutions to propose here. What we can do is select some ideas that seem important to us. First, we can be guided by the three criteria that characterize good science proposed by Philip Kitcher, which we can simplify as follows:

Verifiability. The research for which funding is requested must be independent of auxiliary hypotheses (Kitcher speaks of *independent testability*). For example, "the existence of the planet Neptune is independent of the anomalies in the orbit of the planet Uranus."

Unification. A strategy for solving a problem is important if its applicability is broad, i.e., if it solves a wide range of problems.

Fertility/Productivity. A scientific result is important if it opens new research directions; in other words, if, by solving a problem, we automatically end up formulating many new questions.

To these somewhat obvious criteria, we must necessarily add others that measure the benefit in even more concrete terms. Without a specific order, here is an essential list:

Feasibility. Projects that make too grandiose promises, centered on false expectations or with an implausible timeline, have a low probability of success. This does not mean that people should be encouraged to think small. However, it does mean that the request for resources for a project with a low probability of success should also highlight both the value obtained upon project completion and, in case of failure, what partial results can be used in other projects, even if led by different people. This concept of feasibility indicates the importance of creating value in any case, regardless of success as measured by the parameters associated with the project itself. Any funding request should include an analysis of the impact of research based on its feasibility.

Budget. A project's budget that is too high or insufficient does not imply a direct rejection but requires further analysis. To draw a parallel for generating ideas, consider a startup that has just finished raising capital: it needs to spend on the business's needs, not just because it has funds to spend; it needs to keep an eye on liquidity to ensure the startup has resources available for a sufficiently long window, leading to raising new resources externally or directly achieving self-sufficiency (*runway*). In an academic environment, a research project almost always has a predetermined time window. In other words, a funding request is made for a specific number of years. At the end of the project, an evaluation of the results is made: if the results proposed in the initial request have not been achieved, then it is likely that the direction will change, or more funds will be requested. But if it is already difficult in science to predict results, how can we determine how much time is really needed to obtain them? Hence the idea that, at least in some contexts, the budget for a research project should be interpreted as the runway for a startup, even though it clearly does not aim for financial self-sufficiency unless those working in research create a commercial vehicle to separate themselves from the university. Alongside this idea is the notion that, with the planned budget, we should be able to explore different paths. In short, the budget is something fluid and subject to discussion, whereas in academia, it is often a fixed number obtained as a sum of salaries (what is called OpEx, or operational expenditure, in business economics, and is usually the most significant expense) and laboratory materials (what is called CapEx, or capital expenditure).

Research Context. Proposing to do experimental physics research at a university without a physics department or lacking a laboratory infrastructure brings us back to the criterion of feasibility. However, the risk of respecting feasibility automatically leads to more support for major universities at the expense of less well-known and prestigious centers. Sometimes a department wants to grow a specific field based on enrollment statistics or because it is "trendy." For example, suppose the number of new enrollments in the computer science course grows because artificial intelligence is on everyone's lips, and, consequently, the media talks about easy job opportunities. The department will have to open new positions for teachers, usually also involved in research activities. It almost seems as if the department must respond to market demand. There will thus be a marketing component to attract both students and teachers. Therefore, the context of growth and the desire to grow must also be considered when allocating funds.

Impact. This is obviously a concept that admits various interpretations: how long should this notorious impact take to occur? How should it be measured? Is it just about producing scientific papers, technological spinoffs, the extent to which a paper changes the practice of a scientific community, or the life of a society? Giving too much emphasis to impact, however it is conceived, risks diminishing and negatively influencing the creative part. If, on the other hand, we limit impact to the academic context itself, then there is a possibility of isolating it or at least discussing it. The standard example we have already addressed in the past is a conjecture in mathematics: its importance is determined by the consequences of its proof. In mathematics, it is then reasonable to talk about research with more or less profound impact based on the consequences in the specific field. A rough approach could even be based on questions posed to the community itself, for example, by conducting a census of the most important problems—which can be such for many different reasons, including, simply, a trend if the media (generalist or specialized) have talked about a particular problem more than others. From this census, we could then try to rank problems, without, however, avoiding the difficulties of interpretation and calibration that would undoubtedly arise.

Merit. How is it decided whether a proposal or a research group (or perhaps an individual) deserves funding more than another proposal or another group (or individual)? How much importance should be given to the so-called track record, i.e., the curriculum and results achieved with previous research funds? How do we keep biases in check? What parameters should be analyzed? In his provocative *The Tyranny of Merit: What's Become of the Common Good?*, Michael Sandel (2020) observes that the distribution of economic and social rewards (for example, in terms of hierarchies), even if based on meritocracy, leads to inequalities and potentially harmful marginalization phenomena for the common good. Among the unintended consequences of a purely meritocratic approach, at least in the academic context, Sandel cites the promotion of a hyper-competitive culture centered on individualism; a culture that damages social cohesion and the sense of belonging, with long-term negative outcomes such as, for example, the dissolution of productive research groups or academic stardom.

Inclusivity. This is a central issue. Although the geographical origin of researchers has changed and is changing, much remains to be done on the front of gender and demographic diversity, but also cognitive, generational, family commitments and schedules, physical and mental abilities. Added to this is the fact that most research is concentrated in a few places, with a tendency toward centralization. Inclusivity is undoubtedly an important aspect to strengthen social justice, but at the same time, it fosters innovation and plays a key role in decision-making processes.

Sustainability. Achieving results is crucial. But the sustainability of institutions and careers, in the broadest sense of the term, is perhaps even more so. We will return to this concept in more depth in the next chapter.

Communication plan for results. Besides scientific papers, how will the research be shared with other research groups and society? It is common to underestimate the importance of scientific communication that extends beyond the community itself. Doing the most to communicate the value of one's results should not be relegated solely to professionals who do outreach: as we suggest later in Chapter 4, it should instead be seen as a duty of scientists, to be carried out in collaboration with communication experts.

What we have discussed so far applies to an ideal vision of science. But we must also remember that our goal is to "go behind the actual truth of the matter" (as warned by the opening passage of chapter 15 of Niccolò Machiavelli's *The Prince*), "that he who abandons what is done for what ought to be done, learns rather his ruin than his preservation." Let us then also consider the empirical aspects of doing science.

First, both for those applying for funding and those evaluating it, considering a long list of ideals requires a lot of time and energy. According to a Scientific American editorial (2011), in 2007, in the United States, academic staff spent about 40% of their time writing grant applications. This demonstrates clear inefficiency.

Moreover, since everyone wants to demonstrate success, both agencies and academic staff, the process of evaluating results is superficial and tends not to assign responsibility for any failures. It is true that the evaluation of projects is sometimes based on bibliometrics without observing whether the project has actually achieved the results hoped for and promised during the application phase. What matters is that it has led to something: if an agency has funded a project, in case of failure, in a sense, it loses face.

Empirical considerations are not limited to these just mentioned. John Ioannidis, director of the Meta-Research Innovation Center at Stanford University, has proposed a list of ten problems (mainly empirical in nature) and potential advice for solving them (2018). We believe it is a sensible and content-rich list: for this reason, we will present it in schematic form, even though we do not agree with all its details.

Problem 1. The majority of funds are concentrated in the hands of a few people. Potential solutions: using lotteries to allocate funds. A limit for each individual. Shifting funds from senior to junior researchers.

Problem 2. There is too much opacity in science. In particular, the reproducibility problem needs to be solved. Potential solutions: sharing one's techniques and data, even if keeping this information confidential, can provide a competitive advantage. How? By adopting transparency as a requirement for receiving funding, hiring, and promotions.

Problem 3. The replication of experiments is not sufficiently encouraged. Potential solutions: funding agencies should also support studies that attempt to replicate results, and the evaluation of researchers should take into account not only discoveries but also the existence of confirmations.

Problem 4. We do not support young researchers. Potential solutions: a larger percentage of funds should be allocated to young people, while universities should encourage the hiring of young researchers. (Incidentally, this is a point we disagree with, as it risks being discriminatory.)

Problem 5. Funding agencies are often not neutral. Potential solutions: limiting or preventing funding with evident conflicts of interest; if the conflicts are less severe, requiring transparency.

Problem 6. We support the wrong fields. Potential solutions: monitoring the impartiality of funding in areas that receive more resources; allocating a portion of funds to financing new fields; encouraging researchers to explore new paths.

Problem 7. We do not spend enough funds on research. Potential solutions: improving scientific communication, highlighting efforts to improve and monitor how science is done.

Problem 8. We reward spendthrifts. Potential solutions: favoring funding schemes that reward quality and social value, as well as favoring lean proposals that do not foresee expenses beyond the production of new ideas.

Problem 9. We do not fund risky ideas. Potential solutions: implementing funding schemes that reward researchers leading the research, giving them the freedom to decide how to develop it; effectively communicating that doing science is a cumulative effort and that it is entirely normal for most projects not to yield results.

Problem 10. We do not have good data on science. Potential solutions: funding metascience.

Some of these issues were also reiterated to us by Aram Harrow, already mentioned earlier. We explicitly asked him what concerned him most about how we do science today:

> One thing I would like to change—and it's an idea that Michael Nielsen also supports—is to increase experimentation with the systems that do science. [...] How do we allocate funds? Who is inventing innovative research topics? These are choices that are partly top-down and partly bottom-up. There are influences from industry and governments, but it's also true that research fields follow their own logic. And there are many trade-offs. It's not clear what the best way to do things is. What I would like to see is more diversity in how we do them. And I would like to see scientists testing different systems. For example, we should fund individual projects that seem good, then let the researchers do what they want for 5–10 years. Should we force people to achieve certain goals? I would like to see people trying to systematically understand the benefits of this approach. [...] The problem is that we don't see competing models. To some extent, there are funding agencies that overlap in their interests—and that's good. But there are different approaches driven by responses to industry pressures, for example. A company, rather than the military, might have a different incentive to fund

> one research over another. This does not seem to be based on tests and measurements of progress but on local and accidental decisions. In short, I am convinced that, overall, science is progressing, but we could do a better job by testing the systems we already have.

The funding request process puts individual researchers and research groups in competition. The Arc Institute, inaugurated in 2021, is a non-profit research organization established in collaboration with various California universities and private companies. It aims to study central problems in biomedical sciences by allocating funds for eight years without requiring specific results. This is an approach that the MacArthur Foundation has adopted for many decades.

Initiatives like this have the opportunity to give researchers complete carte blanche: people can focus on creative work rather than trying to achieve (in a short time) results that at least partly align with the promises made in the funding request, generating a mechanism that limits the freedom to follow the most promising path (see for instance Piper 2021).

Another phenomenon to avoid is the so-called bottlenecks in the selection processes of those participating in scientific research. We discussed this with Adam Marblestone, who leads Convergent Research, a non-profit organization that funds research. Regarding bottlenecks, Adam considers a spectrum of types. In the middle of this spectrum, it is easier to solve the problem, but the extremes are complex. Convergent Research's system, according to its director, works well for startups, a term used as a metaphor for the work of small research groups. A scientific startup is a small group (five to ten people) of students and postdocs working with a professor. This type of group can make progress relatively predictably over a few years.

However, the system works significantly less well at the two extremes of the spectrum. On one side, if we have a large group of 12 engineers who need to work on a scientific project, then it is complicated to build a common goal. Obviously, we add, this consideration should not be taken in absolute terms: it is not said that this difficulty is encountered everywhere; it may not exist in every nation and culture, just to give an example.

At the opposite extreme, when venturing into completely new fields, problems arise; a single person cannot, for example, think of doing a PhD thesis based on totally new ideas that have not yet been explicitly formulated. Here, even before doing science, coordination is needed: synergies must be created that constitute the favorable conditions for doing it.

So, to recap, coordinating the engineering effort is difficult. And as an individual, it is challenging to obtain grants. At the individual level, the system can be managed better. At an aggregate level, there are many scientists and many universities: in an aggregate project, many things can be done, but only on fairly clear and well-defined ideas.

In partial contrast, speaking with Ginestra Bianconi, the idea emerged that academia should reaffirm its role in trying to address fundamental problems that impact society as a whole, rather than pushing the interests of specific industrial sectors and allocating financial resources to applications. She then told us that, during an event, she heard someone—awarded significant recognition for their research—complains about the way funds are allocated in the United Kingdom, demonstrating how the problems in science funding exist at every level!

According to Ginestra, even minimal funding across the board should be supported because research funding observes the phenomenon of the rich getting richer, a phenomenon that harms scientific research. For example, today, even publishing in open-access scientific journals requires funds, and there are universities that explicitly want to publish in this way. Many people are unable to publish in high-level journals simply because there is no money. Furthermore, according to Ginestra, funds should be allocated based on the work done in the past: if someone has worked well in the past five years, then they should have a higher probability of obtaining funding. As the director of the *Journal of Physics: Complexity*, published by the Institute of Physics (IoP), Ginestra is indeed using a model based on conventions rather than fixed costs per article (published or read) with various universities. Later, the IoP "reinvests" this money in the scientific community—for example, through a fund that covers the expenses of attending conferences at the beginning of scientific careers.

A moderately optimistic perspective was also offered by physicist James Whitfield, a physics professor in Dartmouth College:

> I believe that science and technology are still growing strongly, and we, as individuals, can do our part. Throughout history, there have been numerous capable people who have pushed the frontiers of human knowledge: we can learn from their examples. From Harvard's endowment [the mobile and immobile capital available to the institution], to the prestige of Oxford, to the past glory of the Library of Alexandria, we have found various ways to concentrate talent and resources. I remain optimistic about the future of

> science and technology, even though their prospects are often presented pessimistically. Governments and institutions can improve the function of science at the local level by increasing support, increasing funding, and removing administrative and logistical barriers. Specific policy choices influence the flow of intellectual talent away from and to a country. Similar to what happens on a social media platform, intellectual talent will concentrate or dissipate in a given place depending on more or less favorable conditions. One of my biggest concerns is how science and technology affect people who did not create it. Digital dependency, deepfakes, the narrative around artificial intelligence, just to name a few, are real problems that we scientists must be careful not to exacerbate. History has shown that technological advances have led society, depending on the case, to progress as well as to war. As human beings, we are destined to invent for creation and to invent for destruction; these two efforts have barely marginal differences, it should be noted. If this hypothesis is correct, perhaps then there is a fundamental limit to the safety with which we, as human beings, can do science and create technology. The step between invention and destruction is short. Today more than ever.

James's words clearly and directly raise a central problem for the theme of this book: we cannot escape the destructive dangers that accompany every scientific innovation, no matter how much it is aimed at construction and not destruction. Thus, concluding this review of how to rethink the scientific enterprise, the main idea that seems to emerge is that we have a *scalability problem*, namely, how to grow scientific knowledge in a way that is not only more effective but also reliable or, as we will say in the next chapter, sustainable. How to address it?

Gathering insights that have emerged throughout these first three chapters, we believe that a solution may come precisely from the union of science and technology, that is, from the machines we are beginning to be able to build today to help scientists in their work, both as producers of knowledge and as evaluators of others' research. Indeed, among those reading this, there are certainly those who have already identified this perspective as one of the directions that has started to emerge insistently since 2023, in parallel with the development of new trends in the field of machine learning, such as the so-called foundation models: programs that use large amounts of data, with billions and billions of parameters, to learn to perform certain tasks without requiring human intervention. The applications range from automatic translation to generating images from words, to systems that propose lines and lines of computer code in real-time based on existing material and user feedback. In short, how can we use machines to accelerate scientific progress? This question will occupy us in the next chapter.

CHAPTER 4

Machines for Science

Notes on the Scalability and Sustainability of the Scientific and Technological Enterprise

At the end of Chapter 3, we identified scalability as a key concept through which to rethink the scientific and technological enterprise. We are witnessing the emergence of a new phase in the relationship between science and technology, one in which we can build machines that decisively impact the speed and complexity of scientific knowledge advancement. However, scalable knowledge raises a series of questions and concerns, which can be summarized by the need to make this knowledge "sustainable," that is, maintainable and effective over time.

Scalability and *sustainability* are the key attributes we will address in this chapter. They encapsulate the most urgent message of this book: identifying strategies to ensure that scientific research can incorporate both. The attributes are closely linked. "Scaling" stands for increasing the volume of results and their depth, as well as the volume of experimental data acquired. In turn, envisaging a lack of sustainability becomes a limitation of scalability. Scaling in a sustainable way requires innovating and optimizing processes, infrastructure, and physical resources, while not neglecting ethical and social aspects, thus meeting the needs of the scientific community of today and tomorrow, promoting its prosperity and its ability to deliver results.

DOI: 10.1201/9781003676614-5

The perspective we will put forward in this chapter is pragmatic. In keeping with the rest of the approach adopted in this book, rather than venturing into the theoretical nuances of scalability and sustainability, we will start from the field: through the voices of many colleagues, we will put forward a sort of executive summary and state of the art on how to think of scalability in science at this singular historical moment. We have divided the discussion into two parts. In the first, we provide an overview of the most interesting examples and examine some possible strategies for making the scientific and technological enterprise scalable. In the second, through numerous interviews, we review six value-laden attributes and processes that are key to the sustainability of this change. This list is obviously not exhaustive, but we hope that it can offer a substantial enough start to a conversation that is yet to come.

4.1 SCALABILITY OF SCIENCE

Retracing the path we embarked on in Chapter 1, one of the fields that today promises the most immediate application of machines that perform, in relative autonomy or at least in participation, the work previously done by scientists is certainly mathematics. We will therefore start with this field to discuss in more detail the use of machines for science.

4.1.1 Technology in Mathematics

Mathematics relentlessly advances into realms of greater complexity, prompting us to harness computational power to aid in this journey. By employing computers and formal languages, we strive to generate proofs for conjectures, thereby accelerating discoveries that might otherwise remain hidden within the vast expanse of mathematical possibilities. Proving theorems is not just a matter of moving the pieces of a puzzle on a surface: even the mechanisms and tools for solving the puzzle are often unknown and need to be invented.

The application of computational methods to prove theorems forms the field of automated theorem proving. With automated theorem proving, we aim to generate proofs from a conjecture and a set of knowledge expressed through a formal language. The central problem of automated theorem proving is the time (essentially seen as a computational resource) required to find the right sequence of symbols to use as a proof. Proving theorems automatically is, in short, like searching for a needle in a haystack: the needle is the sequence of symbols to be found, while the haystack is a gigantic database of mathematical knowledge, symbols, and cases to be computed one by one. The solution lies in the ability to engineer heuristics (shortcuts) to prove theorems. However, before we engage in developing this ability, which can be done through techniques related to the so-called deep learning in artificial intelligence (AI), attention must be paid to the formalization of problems.

The literature on this topic is not scarce, but it testifies to sporadic and discontinuous attempts. A relevant example is the Mizar Project, an initiative launched in 1973 that lasted several decades, attempting to create a large database that could collect much of modern mathematics. The project's objectives were redefined in the mid-1990s based on the perspective of the QED Manifesto (drafted in 1993 mainly at the initiative of computer scientist Robert Boyer) and consisted of creating a database of theorems all rigorously verified by machines (machine checked). By 2012, the Mizar Mathematical Library contained 52,000 theorems. This superficially significant result was only a drop in the ocean of mathematical results, not to mention that it was a database with limited applicability. The project was eventually abandoned, mainly for two reasons (Wiedijk 2007): first, the scarcity of people directly working on the formalization of mathematics; second, the fact that the mathematics understandable to computers did not resemble the real mathematics with which humans work, and the expressive power of automated systems was still too low at the time.

A further step was taken with the development of Metamath (Megill and Wheeler 2019), a programming language that can express theorems and their proofs verifiable by a machine. The Metamath Proof Explorer (MPE) webpage now contains over 23,000 proofs and is a commendable attempt to standardize mathematical knowledge. The project, inspired by Alfred Whitehead and Bertrand Russell's *Principia Mathematica*, requires that the starting point for building proofs be the axioms of Zermelo-Fraenkel set theory. The approach is close to a formalist theory (see, for instance, Shapiro 2000, chapter 6), where mathematics is seen as a game, somewhat like a chess game: there are pieces (numbers and other mathematical entities) and rules (those of the mathematical system in which we operate), and that's all. Mathematics does not express truths, as neither the pieces nor the rules have intrinsic meaning. The problem with the MPE project was the lack of community growth, similar to the Mizar Project; for example, it was not adopted by high-profile mathematicians and mainly attracted the attention of those engaged in mathematics for recreational purposes.

After this brief (and incomplete) historical overview, we come to the most recent and interesting development in efforts to intertwine mathematics and technology to generate proofs: Lean. Lean is an integrated development environment for mathematics based on formal methods, a field of computer science that uses logic to verify the correct (and expected) functioning of software. Lean helps write mathematical expressions in a language that can be checked and manipulated by computers. It helps prove theorems and ensure they are correct. And it allows collaboration with other mathematicians.

By checking the correctness of a proof, Lean goes beyond the bottleneck of trust. "I can delegate to you the task of proving a particular theorem that is useful to me, and the machine will check whether your proof is correct or not. You and I only need to agree on the definitions," explained Leonardo de Moura, one of the founders of the Lean project and its principal developer.

> Among humans, it is really rare to find a project where many mathematical minds work together because there is a trust issue: it takes too much time to check the proofs obtained during collaboration, and as a result, collaborating is difficult. The same thing happens in the interaction between humans and machines: if today a chatbot told you it had proved the Riemann Hypothesis, one of the most important and difficult unsolved problems in mathematics, would you trust it?

Proofs generated by machines would likely be enormous and would require an insane amount of time to check manually. A tool like Lean solves these problems in both human-human and human-machine interactions. As Terence Tao observed: "The [Lean] system allows you to know whether a theorem has been proved or not. Having a gigantic mathematical library with the main results to be treated as data would be like stepping on the gas in research." With the development of tools like Lean, ideally, we will no longer find ourselves in awkward situations such as when an article's author forgets to mention a presupposition or assumes a hypothesis is obvious, and the reader cannot understand how the result was reached and is forced to trust. In Leonardo's words:

> Lean not only helps with correctness but also allows you to browse results because in Lean, everything is defined, and nothing is implicit. It is a tool that helps navigate a sea of information. And for mathematical information, this is invaluable.

As an example of a recent application, we can mention the case of mathematician Thomas Bloom who, in 2022, solved a problem formulated by

Paul Erdös and Ronald Graham. Bloom and Bhavik Mehta were able to verify the correctness of the proof through Lean. Cambridge mathematician Timothy Gowers, from his Twitter account (now X), succinctly highlighted the significance of this episode: "I think it is the first time that a serious result in the 'mainstream' of mathematics does not need to be checked by a human reviewer because it has been formally verified by a computer."

There are, of course, also difficulties in developing a tool like Lean. We asked Leonardo to indicate the main obstacles in continuing the formalization of mathematics on the path taken so far:

> At this moment, one obstacle is the fear of change. There is cultural resistance. The *liquid tensor*, an initiative led by Peter Scholze (GitHub 2022), pushes in the right direction. Many projects in mathematics will end up being like construction sites; the complexity will be so high that it will become impossible to describe a result or problem informally without getting lost. We need machines to help make progress. And the answer to this is not artificial intelligence alone, but formal verification. Artificial intelligence may prove something, but this something must be verified by humans. And if the verification cannot be performed with a simple algorithm, the proof becomes useless. As a result, people will get lost in very long proofs that no one understands. There is then a second obstacle: the difficulty of formalization. You can explain many things in half an hour, but formalizing them may take much longer.

Lean is an example that brings us face to face with the prospect of a digital revolution in mathematics, whose importance can be condensed into the following three points. First, the process opens the door not only to the automatic discovery or formulation of theorems in mathematics (a field in which we are still at an embryonic stage) but also to their automatic verification, as mentioned above. Second, the revolution is leading to the construction of databases that integrate knowledge. Third, as Leonardo de Moura's testimony confirms, the revolution can be based on principles that foster collaboration and trust.

Of course, the field of mathematics is just one of many in the sciences. But it is one that could pave a possible trajectory for other fields as well. To elaborate further on future scenarios of some scientific fields, David Donoho's article "Data Science at the Singularity" (2024) offers a powerful framework. Donoho argues that the recent acceleration in AI research stems from a *Reproducibility Singularity*. This is the moment when the effort to reproduce computational research drops to nearly zero. It's a transformation driven by the maturation of three data science principles into "frictionless," open services: the widespread *sharing of data*, the *open sharing of executable code*, and the adoption of *competitive challenges to benchmark progress* (with public leaderboards). When a research community fully commits to this triad (where "code" can be broadly interpreted as "workflow"), it creates what Donoho calls a *Frictionless Research Exchange*. The case study of Lean suggests that, within mathematics, we may not be too far from the singularity envisaged by Donoho. Importantly, adopting the three principles of radical transparency and reproducibility here sketched, is not just a technical choice, but a strategic one that empowers individual researchers and the broader scientific community to scale discovery collectively. We shall return to this latter topic later on in this chapter as well as in Chapter 5.

Taking a step back. At the heart of our discussion of what doing science means is a direction of development and, linked to it, a future scenario in which machines produce science, guided (perhaps, in some way) by humans. We are not talking about situations that may still be science fiction where robots handle the entire cycle of discovery, from conjecture to proof. However, we are talking about a use of machines that goes beyond computation as a tool for making examples used to support the discovery process. So, doing science is a procedure; if we introduce a machine whose purpose is to carry out that procedure, then we will have a technology for doing science. Is the hierarchy perhaps reversed in this scenario? Science is no longer generating technology, but the opposite is happening—a topic that brings us back to the themes discussed in the first two chapters.

Several projects are already underway that follow precisely the direction we just outlined. We will cite two, starting with DARPA, an organization of the US Department of Defense created by President Eisenhower in 1958 to invest in developing technologies that can play a role in national security. Today, DARPA funds over 250 research projects in fields ranging from robotics to biotechnology, collaborating with industries and universities. DARPA is also famous for developing ARPANET, the computer network that can be interpreted as a prototype of what we now call the Internet (Atherton 2023). In 2023, DARPA opened the Foundation Models for Scientific Discoveries (FoundSci, DARPA-PA-23-04-01) call for proposals to request funds for building, developing, and proving a so-called "autonomous scientist." Ideally, approved projects should autonomously adopt an inductive bias, i.e., choose specific hypotheses to make predictions without human intervention.

Moving on to the second project, in 2023, the non-profit initiative Future House was announced, aiming to build, within ten years, an AI Scientist: "a system that can formulate hypotheses, plan experiments, reason mechanistically about the world, and apply these skills to real-world problems in biology" (Future House 2023). The field of study will be biology because, as the announcement describes, "biology is the science most likely to advance humanity in the coming decades, through its impact on medicine, food security, and climate." Future House's goal is to enable the execution of a hundred times more experiments than can be done today, thereby supporting the scalability of the discovery process.

Future House bases its confidence on new computational tools like ChemCrow (Bran et al. 2023), an agent that uses language models to perform useful actions to generate new knowledge in chemistry. ChemCrow has been able, among other things, to invent a new chromophore, a group of atoms that gives a substance its color. Future House, as a whole, is not only interesting in the context of metascience as a scientific initiative outside the university but also in the methods it proposes. For example, it plans to measure research progress with a Challenge Book, a collection of specific results that the AI Scientist must be able to achieve. Incidentally, the Challenge Book is a model that could be reused, potentially in more general contexts, to push research in specific directions: it has the advantage of setting specific goals while leaving open the direction to achieve them (it should be noted, however, that its potential drawback is that it partly limits creativity).

These are fascinating but still developing perspectives. We turned to Giuseppe Carleo, a researcher at the Swiss Federal Institute of Technology in Lausanne (EPFL), where he runs the Computational Quantum Science Laboratory, for insights into the possible future implications. Regarding the impact of machine learning on science, with a specific emphasis on physics, Giuseppe Carleo points out that, even recently, the most media-relevant results continue to be those on predicting protein structures. The three-dimensional structure of proteins determines properties useful in pharmacology. It is a traditional optimization problem. It is difficult to solve, but optimal configurations can be found automatically. There are also important applications in plasma physics (gases of highly charged particles) and materials science, a field in which machine learning is becoming a tool like any other assimilated by computational physics.

In short, just as the use of computers revolutionized some aspects of physics starting in the mid-twentieth century, we are witnessing a phase in which new techniques are changing how computational science is done. Theory, however, deserves a separate discussion. From Carleo's perspective, a machine remains (for now) primarily a tool for performing calculations: the road to making it a theoretical tool is still long. The key phrase is to simulate phenomena with greater precision and ability than is possible with techniques that do not use machine learning, to explore new things much faster than is traditionally done (e.g., in many-body systems in condensed matter). But we can also imagine a codeless future where programming is no longer necessary, and we use assistants: instead of writing lines and lines of programming code, we will ask the machine for an algorithm to solve a certain problem, and the machine will provide it. If we reach this point, the impact on physics will be enormous: all those scientists that today cannot program because they do not have sufficient familiarity with the required technical language, will be able to program by using their natural language. In this sense, many more people with a more theoretical mindset will be able to be more applied. They may not develop new theories, but they will have a new research tool at their disposal. A tool, moreover, that is easy and intuitive to use.

Beyond the practical changes, Carleo believes that this "new scientific revolution" will almost weaken the concept of truth. He points out that even today, truth in physics is weaker than truth in mathematics, a point we have addressed elsewhere in this essay: the world of physics is given to us, and we must interpret it; equations are almost correlations between different quantities defined experimentally; instead, the world of mathematics is defined, controlled. We report the last part of the conversation with Giuseppe in unedited form:

> *Question*: Isn't weakening the concept of truth, in your view, also a bit like abandoning or distancing oneself from the idea that science can succeed?
>
> *Answer*: Yes and no. A reductionist core remains: we have complex systems and want to understand what governs them. But the experiment will be replaced by simulations, which allow us to better read the systems. Simulating complexity is important, and to do that, you don't need formulas. This is the difference between mathematics and physics: experimental discoveries sometimes change the theory.

A new set of problems emerges when simulation becomes important, and the experiment, understood in the traditional sense, takes a back seat, or at least becomes something more costly and, therefore,

> something to approach more sparingly. The first problem is collecting data on which to base simulation choices when the latter is not entirely dictated by theory, i.e., when simulation is not merely the execution of numerical examples obtained through the specification of variables of closed formulas. The temptation to use automatically generated, synthetic data to save time and resources can lead to contaminating what should be associated with physical reality, with consequences that are difficult to predict, but which we can certainly include in the creation of a semi-true physics. We lose touch with the world—and thus with the sense of science as we have expressed it in the first two chapters—when its expressions end up having a purely syntactic and linguistic origin. At that point, dystopically, a form of non-scientific knowledge is generated, even if logically consistent. It is not excluded that theories developed in this way are still useful, even if they lack the actionable certainty guaranteed by the description of reality.

With these considerations by Carleo, we bring to a close a first set of considerations on how technological advancements are impacting the scientific enterprise. It is important to note that our conversations were primarily focused on mathematical knowledge. Other fields concerned with primary resources for individuals and societies, such as medicine and food, are far from establishing clear paths for utilizing machines in scaling up scientific knowledge. Difficulty in replicating results, misaligned scientific languages, and lack of clearly identified problems to collectively resolve are some of the obstacles to achieve scalability. Far from discouraging us, these challenges should however suggest putting our best efforts into achieving important breakthroughs in those fields in the years ahead.

4.1.2 Science in a Graph?

In the previous section, mathematics has offered us the cue to rethink the collaboration between machines and scientists in the scientific enterprise. A luring but still utopian prospect emerging from the discussion is to represent mathematical knowledge in a graph, or at least in an evolving corpus that simplifies the relationship between the discovery process and its dissemination, going beyond the scientific article. One of the major challenges of such a scenario, however, lies in being able to reconstruct the evolution of scientific knowledge, even if confined to a single field (e.g., mathematics), in a way that is readable even by a machine. Let us try to articulate this solution, attempting to propose a tool that promotes scalability.

The relationships between exprcssions of a language, particularly in scientific ones, have been carefully studied, we can say, forever. Already Porphyry of Tyre, a philosopher who lived in the second-century AD and was known for being a student of Plotinus and editing his most famous work (the Enneads), left us an example to bear witness. Porphyry depicted everything that was known about the world through "trees" whose "branches" expressed fundamental semantic relationships, such as truth or belonging to the same natural kind. In terms of organizing knowledge, Porphyry of Tyre knew his stuff.

Setting aside the fascinating history of "tree" or "network" representations of knowledge, it is important to note the emergence of fields of science that specifically deal with studying scientific knowledge through the analysis of *knowledge networks* or *knowledge graphs* (for an overview, see Ilievski et al. 2020). In Chapter 1, we reported an interview with Valeria de Paiva, who offered us an example: even at the risk of going against the grain, her analysis seeks to enhance the role of informal (natural) languages in representing connections between scientific results (Collard et al. 2022). However, various problems arise. What should be placed at the vertices of the graphs? Scientific results? Which language should be used? How should the vertices be connected?

An important approach is based on the so-called network science. It employs mathematical-statistical methods initially formulated for physical theories (such as statistical mechanics, for example) to analyze how scientific knowledge evolves. Here we will offer a version that we can consider simplified and idealized of a knowledge network (or graph) that will serve as a heuristic for discussing the relationship between science and technology.

Today, there are databases of scientific articles, summaries, and indexes: zbMATH Open, MathSciNet, arXiv, Social Science Research Network, EconBiz, JStor, PubMed, to list at least some of the most used. These are libraries easier to explore than traditional libraries, even though they are already immense and constantly growing. However, the relationship between scientific results is buried in the pages of articles, whether they are paper or electronic. Certainly, the ability to search through algorithmic methods is deeper and more efficient than manual searching through a card catalog. However, the interdependencies remain obscure and recognizable only to expert eyes already deeply familiar with the subject. This research is a painstaking work, suitable for a machine specifically programmed to discover connections, a machine that goes beyond searching using specific terms, symbols, or phrases. To be successful, this type of application still requires much work from both a scientific and technological perspective. And before even considering such a machine, there needs to be a discussion on how to represent scientific results: there may not be an ideal method, but rather many different methods suitable for performing different tasks. (As implicitly suggested in Chapter 3, the scientific article is a product of culture: it is not necessarily the best method for collecting, cataloging, discussing, and disseminating scientific results.)

Let us now imagine formalizing the dependency or relationship between expressions through a graph (or network), as understood in discrete mathematics and complex systems. Specifically, let us imagine a directed graph where the vertices represent the expressions, and the oriented edges connecting the vertices represent some relationship—for example, implication (an alternative could be the temporal order of discovery; but as we will see shortly, this might be redundant). The oriented edges (or arcs) are often visually represented as arrows connecting the vertices.

A vertex is in the tail of an edge if its verifiability implies the verifiability of the expression associated with the vertex located at the head of the edge. The graph is dynamic and changes over time, both with the addition of vertices and with the addition of edges: in principle, one could archive it, collecting the history of the graph's evolution and thus also returning the temporal relationships between its elements. Equivalent expressions in the graph will be adjacent and connected with edges going in both directions. A scientific discovery corresponds to adding a vertex or an edge. Ideally, all scientific knowledge should be contained in such a graph. A catalog, a relational database that includes all science. Clearly, we are talking about a utopian and somewhat pseudo-scientific graph. Who knows whether a final, or at least stable, graph really exists?

Consider two edges from vertices i to j and from k to j, respectively. These edges indicate that propositions i and k allow the discovery of proposition j. In other words, there is a relationship between the propositions associated with the three vertices. There is some dependency. Or, i and k were discovered before j, if we choose to define the relationship based on time.

This image, by association of ideas, remotely recalls the idea of a *causal network*, as described in some approaches in physics (see, for instance, Surya 2019). Spacetime is discrete, and the events that occur in spacetime are connected by a partial order relation that indicates the causal relationships between events. The topic is entirely unrelated, and we mention it only in passing, but we are interested in the order relation. Scientific discoveries are connected by a partial order relation. The graph grows throughout history with the succession of discoveries. A *partial order relation* is a mathematical concept describing a set where some elements are related in a specific sequence, but not every pair is directly comparable. In the context of science, this means that while certain discoveries must precede others—like needing to understand electricity before inventing the light bulb—not all discoveries are dependent on one another. Indeed, the progression of scientific knowledge forms a complex network rather than a straight line, illustrating how new ideas emerge from multiple paths and how foundational research paves the way for future innovations.

From the perspective of our graph, a scientific discovery is made by accessing the expressions introduced in the graph through a careful selection of the scientific literature, and then expanding the graph with new vertices and the relevant edges. Each addition enriches the graph's complexity, illustrating how science progresses through cumulative efforts. The graph grows not just in size but also in the degree of interconnectedness, showcasing the intricate web of how ideas evolve, influence one another, and lead to further discoveries. By representing scientific knowledge in this way, we highlight the importance of both individual contributions and their relationships within the broader context of scientific advancement.

If science is to progress, then there is a need for a system to make every expression in the graph public immediately after a new discovery. For this reason, there must be streamlined mechanisms to access the scientific literature that constitutes the graph. The scientist's work should be doing science, not spending endless hours scouring an ambiguous scientific literature where finding information is difficult; and this is true even today, despite this literature being almost entirely accessible online. Moreover, the discovery work is the truly important phase, while writing the article is a relatively simpler and more straightforward phase. Can the process of searching for results in the existing literature be improved?

Interesting in this regard is the Undermind project, born from two Massachusetts Institute of Technology doctoral students, Joshua Ramette and Thomas Hartke. One of the most characteristic aspects of research is the complexity, variety, and creativity with which scientists approach problems. The skills required are extremely difficult to make explicit; however, at least in certain respects, if we had machines that assisted or guided us in tackling the problems in question, then we could hope to progress more quickly. Undermind aims to develop a machine that performs the functions of a "search agent" that can understand, find, explain, classify, and build a timeline related to a research problem. To do this, it is crucial that the machine "works" starting from a well-constructed data archive, such as a well-organized archive of scientific articles. The initial prototype of Undermind was developed from the arXiv database, with particular reference to the scientific literature in quantum computation. Undermind is not the only attempt to build automated search agents, but also it is currently one of the most interesting. Among the various innovations we can conjecture, one concerns metascience as well. In fact, projects like Undermind allow us to collect data on how researchers think, how they approach problems, and how they manage information; we thus have very detailed and specific knowledge of aspects that have so far been mostly managed informally. This knowledge can be used to represent this part of research more accurately and perhaps rethink it (including considerations on how to delegate it, in part, to a machine).

Returning to the idea of the graph, we will reiterate that it is a mental experiment. For example, a graph of mathematical knowledge will include all theorems ever proven—all propositions, all mathematical truths. The idea of such a graph, however, can be useful in guiding the evolution of research, especially from the perspective of making it scalable. Initiatives such as the Bourbaki collective in the last century were an indirect and very remote attempt to move toward constructing that graph by organizing the body of mathematical knowledge and teaching its disciplines. The monumental work *Éléments de mathématique* is the key example of this effort (Beaulieu 1999). The collective primarily focused on precise definitions rather than connections (because the latter would be observed by those reading), i.e., defining the vertices of the graph rather than the edges. If we go beyond mathematics, then constructing the graph of knowledge seems even more unattainable. However, reflecting on this idea can be useful because it leads us to create a mental model of how science moves and perhaps also of what is needed to understand its value and speed up its progression.

Thinking about the idea of a knowledge graph highlights the importance of having a global standard for writing mathematical expressions and expressing the value of using computers, a standard useful for exploring the graph and highlighting interesting structures within it. In the absence of a global standard, at least a system for converting mathematical (and, more generally, scientific) languages would be needed. The idea of using forms of AI to help make new scientific discoveries is quite obvious, given that machines can read a database faster than humans and find relationships on larger scales. Furthermore, the graph allows us to define and quantify informally the value of a scientific discovery based on the impact that removing a specific vertex from the graph has on the graph.

Continuing, somewhat playfully, to imagine the functioning of the graph, we will note that removing a vertex can lead to various consequences. First, the subsequent removal of any vertex that can no longer be obtained due to the reduction of knowledge resulting from the initial removal will occur. A path consisting of alternating vertices and directed edges in the same direction could end up being completely deleted if the initial vertex of the path is deleted. As a consequence, another vertex will disappear unless alternative paths connect it to the rest of the graph. Although this may seem like a cumbersome process, it makes it possible to quantify the value of a scientific discovery (based on the consequences that the deletion of the vertex associated with that discovery has on the graph). The greater the consequences, the greater the value of the discovery. Deleting a vertex associated with an important discovery will automatically delete that portion of the graph that, all of a sudden, we can no longer obtain. For example, what happens if we delete the Pythagorean theorem in mathematics? What happens if we delete some physical laws? The answers are determined by the impact that the deletion has on the graph. A criterion for quantifying the consequences remains to be determined. Perhaps there is no single criterion but rather a set of criteria that can grow based on scientific discoveries themselves. We will leave the matter unresolved.

A field of its own is emerging in this metascience context: the field of studying how to quantify the value of a scientific discovery through the study of networks, where the nodes could be scientific papers, just to mention one example. Perhaps this approach would not be worse than the crude bibliometrics used today, where the number of citations in scientific journals is considered.

On the most speculative side, a reasonable definition of the analytical intelligence of a machine that autonomously does science could be associated with finding the shortest path between vertices—in other words, a form of optimizing the route to solve a problem. Creativity, on the other hand, results in a jump between non-adjacent vertices of the graph in which new vertices are proposed as conjectures. The new vertices will be fully integrated into the graph once psychological certainty about them has been acquired (as we saw in Chapter 1, psychological certainty is an important characteristic of some scientific propositions, which complements and sometimes replaces the search for their truth status).

Knowledge graphs are thus an attempt to organize scientific discoveries in a way that goes beyond mere bibliographic collection. However, it should be added that managing knowledge may require steps that go beyond organizing electronic material. Indeed, there are interpretations according to which technical knowledge is divided into two components: explicit and tacit, where tacit knowledge, to be transmitted, requires "intimate interaction" between human beings (Li-Hua 2007).

4.1.3 Technology in a Graph?

Parallel to what we have seen for science, we can also define a graph for technologies. Instead of the hypothetical and utopian knowledge graph, we will now talk about a technology graph. Equally utopian, if we use a bit of healthy realism. In this graph, the vertices are technologies. A directed edge between two vertices indicates that the technology at the head depends on the technology at the tail. Deleting the vertex at the tail causes the technology at the head of the vertex to disappear—unless another path is found in the graph.

By constructing a graph that maps technological developments, we can integrate it with the knowledge graph to form a *multilayer network*, as described by Bianconi (2018). In this multilayer network, one layer represents scientific knowledge—nodes correspond to scientific concepts or discoveries, and edges denote the relationships or dependencies between them; the other layer represents technologies, where nodes symbolize specific technologies, and edges illustrate the interdependencies among them.

The edges connecting the two layers indicate the dependence of specific technologies on particular scientific descriptions of natural phenomena. For example, a technology node representing photovoltaic cells would be linked to scientific knowledge nodes such as quantum mechanics and semiconductor physics. These inter-layer connections highlight how technological advancements are grounded in fundamental scientific principles.

A multilayer network is extremely useful for understanding the dependencies between different technologies and the underlying scientific knowledge. By analyzing the structure of the network, we can gain insights into how the removal or alteration of a specific scientific concept (a node in the knowledge layer) might impact various technologies dependent on it. The approach allows us to informally quantify the value of a scientific discovery based on its influence within the network. Furthermore, such a graph facilitates the identification of critical scientific areas that serve as foundational pillars for multiple technologies. It aids in strategic decision-making regarding research and development by highlighting which scientific domains have the potential to drive significant technological advancements.

In essence, a multilayer network of science and technology provides a systematic framework for exploring and understanding their intricate interplay. Such a network can also aid in identifying critical areas where investment in research could lead to significant technological advancements. For instance, if a particular piece of scientific knowledge is connected to many emerging technologies, then focusing research efforts there could have widespread benefits. Similarly, understanding these connections can help educators and policymakers make informed decisions about which scientific fields to support and develop.

4.1.4 A Science Database

The scenarios—although still hazy and hypothetical—of a science and technology graph seem to suggest a possible approach to the issues of scalability that we saw in Chapter 3. We aren't the first to think this could be a solution; consider the Future House project, which aims to use AI to advance the natural sciences more efficiently. Let's explain how the scientific enterprise could be made scalable by introducing the idea of a science database.

Many studies now observe how controversial bibliometrics is in its value (Baffy et al. 2020) and, more generally, how the journal publishing system is at a crossroads (Krumholz 2015). Publishing in a journal becomes almost a distraction from the research itself, turning into a separate process. There have also been striking episodes of scientists refusing to publish in scientific journals. A notable example that received media attention is Grigori Perelman, winner of the Fields Medal, who never published the articles proving the Poincaré conjecture but instead disseminated them on arXiv (2003).

How to determine the quality of a scientific paper without peer review is a difficult problem to solve. This is where, in the long run, some automation is needed, and where the image of the scientific knowledge graph comes back. Ideally and utopistically, there would be no need to write scientific papers, but simply to add a result to a database fed daily by the work of every individual scientist or research group.

Such a single database would eliminate the need to waste time writing introductions, abstracts, titles, and conclusions—all required to package a result into an article. Let us briefly hypothesize some of its features. The closest thing to this proposal would be a collection of living papers: papers that don't necessarily aim to disseminate information about specific results obtained in a certain timeframe but rather to discuss a problem.

These living papers would be dynamically updated by a potential multitude of authors. In this way, while no less time would be spent writing papers, at least more focus would be placed on the science and the results themselves. However, even with living papers, there would still be a group of reviewers to determine whether a result is correct.

Doing science by entering a result into a database like the one we are hypothesizing would have some likely advantages: in addition to saving time in various phases of scientific production, it would also facilitate maintaining a thread in research, making its evolution more orderly and easily traceable. On the other hand, we can also imagine some open questions and prefigure some issues. For example, which changes would this scenario entail for the communication between scientists? And what governance structures and principles can we envisage for the database?

The scientific knowledge database eliminates the need for articles in specialized journals and, in general, the need for articles. We would thus have an effective way to streamline the process of disseminating science: every time a result is obtained, it would be enough to propose adding the result to the database without publishing an article. Of course, there would still be review articles, compendiums, books, and all the tools for teaching and collecting, through critical thinking, particular results that fall into a specific category. Conferences would also remain. A conference is an important opportunity to share results through seminars and conversations among people.

From here, the path to digitizing science is short, which may also push for peer review to be done by machines, albeit with human participation. A database should therefore have machines capable of analyzing it. It would also be capable—and this may seem like a strong opinion—of performing automatic peer review of papers based on the existing literature in the database and the formal correctness of the result. This is plausible in mathematics, where verification is a logical operation. However, it would be very difficult to determine in less-formal areas of knowledge, such as biological sciences.

Given the significant results in language and inference today, it's not absurd to think that, in the more or less near future, we will have the technical tools to perform automatic peer review of scientific papers even outside of mathematics. Automatic peer review would eliminate human factors associated with the peer review process. We can also conjecture that we'll see a wide dissemination of scientific papers—for example, literature review papers—written directly by machines.

Of course, other types of problems will arise that are difficult to determine at this point. For example, some results will be difficult for a machine to understand because they aren't connected to the existing literature. When defining a new term or concept, it's difficult for a machine to determine the correctness of a result without formal logical inference and analysis of the result itself. Or, regarding articles "signed" by machines, we would need to ensure the result is verified. The first risk is that those who maintain the databases could have excessive power to influence and anticipate the evolution of science. A second risk is viruses or the introduction of false material, even by external actors. A third risk is that it could become easier (perhaps too easy) to exercise a form of censorship on research.

On the other hand, we are not suggesting that machines should be left to their own devices, but rather that they could allow for the "certification" of a scientific paper more accurately and justly than humans. This verification by machines is already happening in the field of computer programming. It's clear that verification has problems related to the computational complexity needed to determine the correctness of a particular expression. However, this can be mitigated based on literature research and the use of both formal and natural languages.

We haven't yet answered the question of who or what entity would be responsible for managing the database. Those who manage the database could take actions that damage it, either voluntarily or involuntarily. External agents with malicious intent would also pose a risk. As in all situations where there is a single point of vulnerability, the database requires broad governance and an engineering approach that leverages distributed systems and replicating information; clearly, it will become necessary to do everything possible to avoid a concentration of power. And these are all aspects for which approaches already exist. Moreover, AI itself poses risks: the introduction of false statements into the database or statements that are true from a logical standpoint, but false because they are based on synthetic and not real experimental data. Managing science at scale is a significant gamble.

Ultimately, the proposed scientific knowledge database offers a compelling solution to the scalability challenges. While its implementation presents significant hurdles, particularly concerning governance, data integrity, and the nuanced application of AI, these are not insurmountable. The transition to such a system would streamline research dissemination, enhance efficiency, and foster a more orderly and *traceable evolution* of scientific knowledge. By thoughtfully addressing these complexities, this model could fundamentally reshape the scientific enterprise, allowing researchers to dedicate more focus to discovery.

Of course, a major benefit of this shift to an open, dynamic database would be the opportunity to share far more than just finished papers. Science is often a process of trial and error, yet the current publishing model only rewards successful results. A comprehensive database would encourage the open sharing of negative results, failed experiments, raw data, detailed workflows, and the specific obstacles encountered. This transparency would allow the community to learn from these setbacks, prevent redundant work, and accelerate progress by building on a more complete and honest picture of the research journey. It would represent a natural evolution toward a truly collaborative and more accountable scientific culture.

4.1.5 Androids as Muses for Humans or Humans as Muses for Androids?

The idea that machines could revolutionize the way science is conducted is relatively recent, but not entirely new. Dashun Wang and Albert Lázló Barabási discuss this concept in the already cited *The Science of Science* (2021: section 4.1), as they introduce Adam, a machine trained to uncover functions for those parts of yeast DNA whose roles remain unknown.

However, we must avoid the mistake of thinking of machines solely as passive tools for taking measurements or implementing algorithms. Machines are used to produce information, acting as agents—not necessarily to be defined as "intelligent," but at least responsible for generating conjectures and hypotheses, traditionally a purely human domain. This potential role could have a significant impact. Additionally, methods are being developed for automated measurement through robotic sensors, allowing for the autonomous collection and classification of much more information.

In the early 2000s, Nobel laureate in physics Frank Wilczek predicted that in a hundred years, the best minds in physics would be machines (2006). The cooperation between humans and machines in science is not inevitable, given current dynamics, but it will likely become natural because it is a necessity. Accelerating the discovery process, communicating results, cleaning data, and repeating complex experiments are all phases of scientific work that welcome automation.

Machines capable of uncovering new scientific understanding are closest to Wilczek's vision. This is likely the most challenging class of machines to build. To address the challenge, we are first of all in need of a map of the process. This is already a topic of dispute among scholars. Just to exemplify, following Stefan Kramer et al. (2023), we can roughly schematize it into six steps:

1. Formulate a scientific question, for example, "Why, if I throw an object upward, does it eventually fall back down?" These questions concern phenomena for which our current understanding is unsatisfactory.
2. Formulate a hypothesis, bearing in mind that access to scientific literature is useful for this.
3. Design an experiment, knowing that any experiment must strengthen or validate the hypothesis; otherwise, it has no value.
4. Conduct the experiment.
5. Analyze the experiment's results; from here, one returns to Step 2, as the results provide information about the hypothesis.
6. Communicate the results.

With a map of the process at hand, we can then start conjecturing what sorts of machines we need to produce new scientific understanding. A relevant effort in this direction is led by Mario Krenn's research team, which has used the term *android* to describe machines employed in scientific understanding. In a 2022 article (Krenn et al. 2022), to the question, "How can androids contribute to new scientific understanding?" the team proposed a classification into three groups.

4.1.5.1 First Group

Androids that act as computational telescopes or tools for measuring the properties of physical systems. These androids perform calculations, such as molecular dynamics simulations. These calculations allow for observations that are difficult, if not impossible, to make through direct scientific experiments, such as when the number of measurements to be made is too large.

4.1.5.2 Second Group

Android muses, usable as a source of inspiration for new ideas that are then generalized by humans. These androids play the role of suggesting new perspectives that fuel human creativity. This could involve discovering regularities and symmetries in data through computational methods or identifying particularly interesting topics in scientific literature that is too vast to be analyzed "by hand," for example, a key step in a mathematical proof within an already published article.

4.1.5.3 Third Group

Androids are capable of scientific understanding and, consequently, ready to replace humans in a part of scientific production. These androids "replace humans in tasks such as generalizing observations and applying scientific concepts to new phenomena." There are two sufficient conditions for these androids to be capable of scientific understanding: they must know how to use a theory in different contexts and communicate with people. The second condition actually presents various problems: What if the formalism is too difficult for humans to understand? What if the interface for interacting with the machine does not allow the full flow of information? It is already complex for humans to understand each other when discussing science in the same language.

Android muses have the potential for great utility because they can algorithmically assemble information that can later be reused by humans, much like giving us the pieces of a puzzle that we then assemble with our own efforts. Dutch theoretical physicist Mikhail Katsnelson, reflecting on the rarity of new ideas and the difficulty we encounter in communicating them, writes:

> Most scientific work is not creative but is merely putting together already known things: if the method proposed in article A is applied to the problem formulated in article B, keeping in mind the following verification on the system described in article C… This is an extremely useful activity because it is what science mainly consists of (in terms of time spent, resources, etc.). […] When Einstein was asked if he was afraid of forgetting his great thoughts, he replied that thoughts come to mind so rarely that it is impossible to forget them. […] Turning an idea into something real, into calculations, experiments, and so on is almost always a problem of exorbitant complexity.

4.1.5.4 Fourth and Fifth Groups

To these three groups of androids, we can add two more. The difficulty of generating ideas leads us to define a fourth group: androids that assist us in amanuensis work. More patient and faster than us, but capable of complex interactions (unlike the calculators we are familiar with), these androids would be attentive listeners and very skilled in reducing the "exorbitant complexity" mentioned by Katsnelson, helping us move from an idea, initially generated by humans, to scientific knowledge obtained in collaboration with the machine, which does most of the "dirty work." Finally, we can imagine a fifth and final group, which includes androids capable of scientific understanding in a situation of total automation.

Today, there is no system with the capabilities of a fifth-group android. Nor is there even a fourth-group android, that is, a machine capable of deciding its own goals. When might we get there? In February 2020, an important workshop was held at the Alan Turing Institute in London to discuss the realization of fifth-group automata by 2050 (Kitano 2021), about 50 years ahead of Wilczek's prediction. The future will tell.

4.1.6 Science on a Hyperscale

Peering through the peephole into the room of science fiction, one notices a future scientific culture based on cooperation between humans and machines.

However, two problems naturally arise from this vision. The first is almost predictable, especially since it is widely discussed today, and concerns the quality of data. A machine can analyze a vast number of data points, many of which are entirely useless for research purposes. To achieve quality, however, costly data-cleaning processes may be necessary, which are capable of providing some guarantee of accuracy, completeness, and so forth.

The second problem concerns the verification of the process that transforms data into valuable knowledge. Without verification, we cannot trust that the computational process truly delivers what it promises. In this science fiction room where machines, in a sense, produce scientific knowledge, humans do play a role as reviewers and controllers, but not so much of the process itself, because it is beyond our capacity to analyze. Our role is entirely in making decisions before the process and, subsequently, in potentially filtering the scientific results. Let's take an example.

Imagine a surreal situation in which machines produce such a large quantity of scientific observations, such as mathematical theorems, that it becomes impossible to determine which of these are genuinely useful. (Already today, if we formalized all the mathematics produced annually on the planet, then we would get over a billion lines of code.) A sea of discoveries, paradoxically, can result in the impossibility of describing or predicting. This situation, if we think about it for a moment, is indeed a fiction but, as Pirandello would say, it can become reality. Thus, we arrive at a science conducted on a hyperscale, where databases are difficult to explore. The result is a projection of the physical world that is as detailed as it is vague, containing mostly unusable results.

The oddity of this hyperscale science would be a new phase in which science no longer consists of studying the world, but of inventing a science of the representation and models of machines. The representation would thus become an inter-layer between physical reality and humans, making us lose contact with the reality we can experience with our senses at the very moment we interface with the formalization of the world provided to us by machines that, let us emphasize, are incapable of distinguishing between what is scientifically relevant and what is not.

4.1.7 Android Creativity and Human Creativity

With the prospects of a science on hyperscale, conducted with and by machines, we have substantially concluded our examination of the scalability of scientific knowledge. Before moving on to consider its sustainability, we should entertain, albeit briefly, one last topic: ingenuity and creativity. Androids do not get distracted. They do not procrastinate. At the same time, however, the only way they interact with the world is through what we humans have provided them, unless we manage to equip them with sensors. To what extent can we regard the output of the execution of a function in a machine as a creative act?

The answer is, in a sense, quite straightforward. When discussing the knowledge graph, we defined creativity as a leap between non-adjacent vertices of the graph (i.e., not connected by a single edge), proposing new vertices as conjectures. Since it is expected that this task could be performed by machines, in this sense we can legitimately speak of creative androids. Let us elaborate on this point.

In general, the complexity of a problem is associated with the space that must be explored to find the solution to the problem—the *search space*. Sometimes one can try to determine the most efficient stratagem to solve it deterministically, but even in this case, methods that include heuristics and intuition are useful. This search space also encompasses untraveled paths. It is the space of the thinkable: an immense space where we wander, driven by experience and our sensory capacity. Thus, creativity goes beyond deterministic combinatorics, moving into a space hosting new problems with choices that have a random flavor, that is, made by chance or guided by unconscious mechanisms. Creative androids, having more computational power than humans, may suggest connections after training in similar search spaces and on large amounts of data. In this sense, we can imagine that creative androids will suggest conjectures that we would not fancy and that seem surprising or novel to us, exceeding the most immediate combinatorial options we could have imagined.

The recent applications in images and videos of generative models have generated considerable interest, both for the quality of their results and for the surprises we did not expect. These are formal procedures that take as input the linguistic description of some image to be generated and synthesize an image "close" to the given description, including random variables, i.e., rolling a die to make random choices that respect the linguistic labels provided as input. The input "Parrot sitting on an elephant," for example, will not yield a specific parrot, but some parrot and some elephant. The model will produce everything the text did not specify by working on a large set of images already paired with words and, essentially, rolling the die every time a new connection between pairs needs to be generated. Generative models can be extended beyond images and can even play a role in the creativity of scientific activity. Methods of inverse design already exist (which start from a desired result and devise strategies to achieve it) that attempt to build chemical structures with specific characteristics. The goal is to produce new substances with certain properties by training the model on a set of known substances (Anstine and Isayev 2023).

Assuming that in the near future the results will improve, also thanks to the impetus of substantial investments, what role remains for human creativity? Isolating the context to mass communication, we discussed this with Luca Lorenzini, founder and creative director of Small, an independent advertising agency.

According to Luca, creative ideas that worked 40 years ago still work today. The most successful advertising campaigns have always been based on profound concepts that work regardless of the medium used to spread them: from print to TV, from billboards to social media. While the foundations that help make a campaign memorable and effective remain the same, the advent of newer, more modern, and technological communication platforms has influenced how an idea comes to life and how it is told. For example, once all videos were shot horizontally; today, with the advent of smartphones, many videos are shot and shared vertically. With the arrival of social media, and due to the bite-sized consumption of content, when sharing a branded video, it is essential to grab attention in the very first seconds to avoid the user getting bored, scrolling, and immediately moving on to the next content.

We then asked Luca to tell us how he and his team use various technologies in the creative process to seek inspiration: for example, they watch videos, search for images, and listen to songs. Technology helps in many different ways, but it usually cannot provide the spark that underlies all the most unexpected and innovative ideas. At present, generative AI can only produce mediocre ideas. For the most part, these are first-level ideas that could have been created by someone who does not do our job. For example, if we ask a machine to come up with ten different banners with ten different calls to action, then the machine can do it. We can then decide to put all ten versions of the banners online and let only the version that achieves the best results survive. What the machine cannot yet do today is imagine unexpected ideas: ideas so wrong that they are just right. Machines, as of now, do not understand the allure of breaking the rules. Luca imagines a future where machine-generated ideas may be indistinguishable from human-generated ones, but our ideas will continue to be more captivating precisely because they were conceived by a human being. It is similar to handmade shoes, which are more attractive and expensive than shoes created by a machine in a factory. This is because they are made with more attention to detail, but it is not just that: behind handmade shoes, there is the work of a person, their craftsmanship, their passion, their precision; there is their entire story. Thus, we can imagine a future where ideas have different origins, human or synthetic, but the most valuable ones will still be those that are human-made.

4.2 SUSTAINABILITY OF SCIENCE

We have outlined the attempts to scale scientific and technological endeavors through the use of machines, pointing to an imminent (potential) revolution: a scenario where machines will (would) be used to manage, at least in part, peer review, funding, experiments, and perhaps even the most creative and seemingly "human" theoretical aspects. This raises an additional, fundamental question of metascientific entrepreneurs: What principles should inspire, inform, and guide this revolution? How can we ensure that we will not get to the point where we think that—to put it bluntly—*it can't go on like this*?

This is a question that pertains also to the so-called "human-centered AI." However, we want to minimize the risk of misunderstanding the future of scientific research as being synonymous with that of AI, as digital technologies should be interpreted primarily as tools, despite their undeniable impact on our perception and modus operandi in doing science. Thus, given that our reflection focuses decidedly, and more broadly, on scientific research rather than on AI, the key term we will use is *sustainability of scientific research.*

"Sustainability" is a polysemic and slippery term. In this context, it is important to understand this term broadly, as a *theory of resources*. Simply put, action *A* is sustainable if we can continue performing *A* without degrading a certain property *P*, for example, our health or that of future generations; action *A* (writing grant proposals, for example) is more sustainable than action *B* if *A* uses fewer resources than *B* to achieve the same result (where "resources" can refer to time, the number of people, or energy). The scale of the duration, of course, depends on the process itself: since "sustainability scientific research" must be intended in reference to the relationship between science and society, we can conjecture that the scale will be a relatively large multiple of a human generation.

Our notion of sustainability is therefore about ensuring the long-term vitality and impact of research. We will explore our topic by closely examining six keywords connected to the idea of sustainable science: phronesis, innovation, communicability, risk, last mile, and ordinariness. For each of them, we turned to one or more experts for insights on how best to approach them, to ensure an expanded dialogue and a plurality of perspectives. What emerges is the need for practices that optimize the use of finite resources and nurture talent so that intellectual capital can flourish over time. Ultimately, the aim is to cultivate scientific endeavors that create the most impact and translate into useful technology.

4.2.1 Phronesis

To accomplish the tasks that scientists perform today, machines must be able to develop similar competencies. But what does it mean to be competent? And how does one become competent? According to a famous article from the early 1990s (Ericsson et al. 1993), it takes long hours of work, an observation that in a sense confirms something intuitively obvious. From this arises the spontaneous rule of 10,000 hours: the magic number to become an expert in something is 10,000, more specifically 10,000 hours of practice (Gladwell 2008). There are human activities where competence is relatively easy to determine. Consider all those contexts where competence is entirely relative to a precise metric: the quality of a golfer is determined by the number of strokes used to finish the set holes; the quality of a sailor is determined by the time taken to complete a regatta. Things get complicated when the metric is difficult to define. For a pilot, competence is associated, for example, with the number of flight hours. And who would we prefer to operate on our hearts? What kind of competencies would we want a machine that advances scientific knowledge in partial autonomy to have?

In ancient Greece, there was a special term to indicate certain forms of competence: *phronesis*. Phronesis is a form of knowledge that combines theoretical and practical knowledge, guiding us in actions. For example, phronesis governs our daily eating habits, choice by choice. When it works properly, phronesis guides us wisely. Evidently, phronesis also plays a role in competence: being competent means knowing how to guide one's actions in a specific field with a certain degree of wisdom.

Ashley Rose Mehlenbacher, a professor of science, health, and technology at the University of Waterloo, Canada, has made phronesis a central idea for understanding contemporary scientific knowledge. We spoke with her about how scientific competence must, of necessity, be defined contextually, and we discussed how complex it is to render all this in terms of machine action. Here, we present the content of our dialogue in prose form.

In her book *On Expertise: Cultivating Character, Goodwill, and Practical Wisdom* (2022), Ashley explores some of the ways we currently think about expertise, arguing that we must do so in several complex ways to avoid reducing it to a mere word devoid of meaning. To fully understand the richness of expertise, we must pay attention to those permutations of thought that experts undergo. When we think of experts, one way to do so is to think of someone who possesses some type of knowledge—where the term "knowledge" is not used in a broad sense but refers to a form of knowledge that most people may not have, and, more importantly, to knowledge that is significant to other people. Experts, then, are consulted by others to help them understand something about the world, their lives, or themselves.

Now, the evaluation of competence is contextual. We saw this throughout the pandemic, when we had to identify certain people as experts to be interviewed on newscasts, for example, and we had to understand what, exactly, these people were experts in. Very often, it is advisable to seek intermediaries for experts. In general, we need an accreditation method, for example, someone to examine these people and determine which of them are experts on the issue at hand (a task we might expect the media to perform). By implementing these processes, we begin to understand who might be an expert; the problem is that it takes quite a bit of work to truly assess competence in the field we are seeking information about.

In her book, Ashley also discusses the ethical dimension of expertise. In short, if an expert is to do something in the world, then they are immediately implicated in ethical issues as well. To assess an expert, we must not only consider how much technical knowledge they possess and what can be technically done with that knowledge, but we must also evaluate how it can be applied to the problems and challenges that other individuals face, in a way that supports their needs. Therefore, we speak of expertise as practical wisdom, where we also see an ethical component. Thinking of experts through this lens helps us move away from the model often used in academia, where a professor is seen as a sage on a stage. We must move away from that idea: experts are genuinely at the service, or at least in conversation, with the needs of a community or group of people (some of whom may themselves be experts).

Another important theme to touch on is the idea of good intentions. What begins with the good will to apply knowledge often ends in the inability to do so due to a lack of strategy. In other words, good intentions are not enough: everyone has good intentions, but to realize them requires a form of practical wisdom that includes, for example, taking a reflective stance on what one is doing and why. Doing so is already a challenging endeavor. In specific cases, such as that of a rhetorician, it seems entirely normal to assume that for someone performing such a task, it is pragmatic to think first and foremost about their audience. But we must do the same with technology. A classic example is virtual reality. Many women have problems with VR headsets because, quite simply, they were not designed for women (they cause, for example, a sense of nausea during use). Thinking about who is included or excluded, what concerns or problems these people might have because we are including or excluding them, is an important step in questioning what the technology is meant to do and how it might reinforce or create new problems for humans. And the very act of allowing oneself to think about who is involved and who is not can truly be the key to reflecting on the perspective from which one is considering the problem.

Looking to the future, we can outline three challenges, especially when considering the hyper-specialization of disciplines. The first is collaboration. There is a strong demand to make research multidisciplinary. But aligning different areas of expertise is very difficult: it takes a lot of work to do something productive and coherent. The second challenge is the public. Here we have a wide range of issues at play. Historically, differences of opinion among experts have led to a loss of trust in particular groups of people or communities. Regaining that trust, once lost, is truly challenging, but it is obviously the kind of work that experts must do. There is also the issue of experts being discredited due to misinformation campaigns and tactics. We see this, for example, with public health experts.

Additionally, the public is also affected by this issue, as they must relate to the supposed experts: how to assess the credibility of a supposed expert, whether it is appropriate to follow the advice this person offers, and what the implications might be for their daily lives. For example, it could simply involve collaborating with one's family doctor or understanding who the expert to listen to on public health issues is when a pandemic is rapidly evolving. Moving forward, a third challenge concerns competence: whom can we trust? And, in general, why should we trust scientists? How can people with certain types of specialized knowledge earn the trust of the individuals they want to work with? It is crucial to emphasize that there is no monolithic and standalone public; rather, there are multiple factions and interest groups.

Finally, we come to the role of technology: how could it help generate trust? In many ways, it currently seems to do the opposite; consider social media and all the platforms that increase the complexity of an issue. This is a largely unexplored topic. We can conjecture that trust is fundamentally a human problem, as any technology we are developing or trying to position as a solution to a problem ultimately depends on how we design it and for whom we design it. And design not only replicates our biases but also our types of relationships in general; all of this is reflected in our trust in that technology, or its lack thereof.

Reflecting on our conversation with Ashley, we were left with some questions that we did not have the opportunity to ask her and to which we will return to in Chapter 5: What if machines were to design technology? More broadly, what if machines were to play the role of experts in the near future? How can we instill the right forms of phronesis in a machine?

4.2.2 Innovation

Technological innovation goes beyond the construction of machines and artifacts. However, its consequences, as well as its evolution, are challenging to gather under a unified framework. We are capable of evaluating some contexts more directly than others; this is, for example, the case with transportation, where being able to move between two cities in a reasonable time not only expands commercial opportunities but also certainly changes our perception of distances, as well as interpersonal relationships. In this area, therefore, a complex feedback loop is created between changes in technology and changes in our behavior. The situation is even more difficult when it comes to areas that have recently grown, such as social media and large language models (LLMs); consider the unintended consequences of interference in election campaigns, isolation and social anxiety, the spread of false information, bullying, potential privacy violations, concentration of power… The list is long.

Defining technology, therefore, can have extremely practical implications. Emily Michiko Morris made this point with the following observation related to industrial property, which, as far as patents are concerned, is centered precisely on the concept of technology:

> The argument concerning what can be patented is a confused area of law that needs clarity. Despite its long history, the U.S. patent system has never been able to provide a precise definition of what 'technology' means in the context of patent law.
>
> *(Morris 2014)*

With sarcasm, Kline points out (1985) that at the end of the twentieth century, the only point on which people agreed regarding technology was its importance; discussing it would be a bit like talking about weather forecasts. Kline himself proposes various uses of the term "technology": (1) artifacts: all objects manufactured by humans; (2) a socio-technical system of artifacts: people, machinery, processes, economics, and politics, and everything needed to produce artifacts; (3) knowledge, technique, or methodology; (4) a socio-technical system for using artifacts: a system consisting of artifacts and people who perform actions for which the use of such artifacts is indispensable.

The complexity of technological development is heightened by rivalries among those working in technology. We are talking about companies competing in the same market or pushing to create a positive perception (such as appearing innovative to customers) that may not even be based on the product they sell. Presumably, nature (and this differs from the practice in technology) is not in competition with scientific research, and it seems that nature has no incentive to make research more difficult. In other words, in scientific research, nature is not our antagonist.

Competition in technology, then, plays out both in the quality and cost of the offering and in the perception of those who use the technology. Defining metrics to compare technological objects, to compare their utility and impact on society, is difficult. Commercial value and success in adoption will be functions that go beyond such metrics. A technological object can indeed have an effect on the self-perception of those who use it, but it is also the perception of the company that produces it that gives the object a connotation and creates potential demand. In deep tech, it is common to see companies working to generate demand for a prototype—and relative mindshare (notoriety)—even when the timeline of its actual market entry is unclear.

While scientists compete for funding by trying to convince organizations, technologists and companies compete to create and capture market shares by convincing customers. The transfer of technology between academia and industry in the form of spin-offs and licensing accelerates the transition between the two realities. Research transforms money into knowledge; "technology transfer turns knowledge into money," quoting Post-It inventor Geoffrey Nicholson. Ideally, this should be a virtuous circle. Reality only partially reflects this. Today, research is often linked to the impulse of funding and seeks to achieve what is perceived as success. Moreover, the resulting technology may not tend toward solving societal problems.

How do machines acting with relative autonomy change the balance and rules of technological innovation? For example, we might have machines handling technology transfer: from evaluating the potential of research, perhaps determining its value, to transforming it into new technology. How would this impact innovation?

This question implicitly raises the issue of defining which aspects of innovation depend on contextual nuances or on an assessment that is essentially made by those who must receive that innovation: in other words, the question of authority in recognizing what counts as innovation arises: can a machine decide on behalf of a group of people or a society?

4.2.3 Communicability

We have already said how important sustainability is and that it needs to be thought of, in general terms, as a theory of resources. But there are other points not to be underestimated. One of these concerns communication. A society based on a close relationship with machines requires a great ability to communicate the value of technology. To whom? And in what ways? As highlighted in the conversation with Ashley Mehlenbacher, we must not make the mistake of thinking that the relationship between science, technology, and society is divided into large monolithic blocks. There is no single public to communicate technology to; rather, there are multiple publics, even among those who do research and those who build technologies. We, therefore, have many communities with different skills and interests, among which effective communication must take place.

Discussing these topics requires a certain open-mindedness, because more than once in academia, we have heard communication spoken of as a mere accessory to research. Those who support this idea have their reasons: for example, they believe that in the long run, it becomes evident what truly creates value regardless of communication, almost as if value was an emergent phenomenon. We have a different opinion and believe that wrong communication can cause enormous damage, for example, by channeling financial resources toward false problems.

The issue needs to be managed with pluralistic skills, taking into account the interests of various publics. This is precisely why we spoke to four people who, while coming from different fields, all see scientific communication as a crucial aspect of their work.

4.2.3.1 Vincenzo Cosenza: Incrementality and Polarization

The first conversation we had was with Vincenzo Cosenza, an expert in marketing and innovation and the author of one of the most well-known Italian blogs on these topics (vincos.it). We asked him what the most difficult challenge in communicating technology is. According to Vincenzo, it is language, or rather, the ability to speak or write in a way that arouses interest even among non-experts. The goal is to try to make innovation understandable in the language of those who use innovation to improve their lives, rather than in the language of those who produce the technology. It is not easy to communicate less-striking innovations when they are brought to market.

A central point, according to Vincenzo, is incrementality. Many innovations are incremental, adding small pieces each time: if we take them as a whole, the innovation is evident, but it is difficult to convey what has changed from one year to the next, or from one model to the next. In fact, often little changes. Perhaps the process technologies and what is behind them change, such as savings for a company or increased performance, but these are things that do not have a significant impact on the user's life; and note that here we are not specifically talking about consumer electronics products. Thus, to engage the consumer, communication sometimes becomes hyperbolic, using, for example, exaggerated adjectives precisely because it struggles to communicate increments whose scope, all in all, is relatively small. Here the risk is of being dishonest; it is better to communicate without hyperbole. On the other hand, it is also difficult to blame anyone because it is society itself that, with its demands, expects a continuous flow of significant new developments.

Another theme not to be overlooked is polarization. Those who communicate in the technological field must succeed in countering a media communication that, on these topics, is very polarized today. What does this mean? At some moments, the media exalt technology and represent it as salvific. At others, they use communication that speaks of technology as a destructive and risky force. Those who communicate must be careful of the oscillation between these two very distant poles: salvation and destruction. This brings us back a bit to the words of James Whitfield encountered in the previous chapter: "we cannot escape the destructive dangers that accompany every scientific innovation, no matter how much it is aimed at construction and not destruction."

4.2.3.2 Aparna Prabhakar: Simplicity

An additional and related observation to those of Vincenzo emerged from the conversation we had with Aparna Prabhakar, Senior Vice President, Strategy and Sustainability at Schneider Electric. According to Aparna, the most important but also the most challenging thing is simplicity, i.e., not complicating things when communicating them. In the relationship between science and technology, science represents the creative moment, and technology the process that makes an innovation potentially usable by many users. However, communicating the value of something to those users is difficult. The difficulty lies in clearly expressing where the value lies. Computers understood as tools for calculations (something not too far from how we interpret them today) were developed during World War II. At that time, however, a 15-year-old could not have used one because it was too complex and too distant from everyday life. It took decades to make computers a technological tool usable by society in an extended sense.

According to Aparna, we can divide the technological innovation process into two parts: the first is structure, which science deals with; the second is non-structure, which technology deals with. The latter seeks to understand the areas of use for which simple narration and communication are essential. Is there any technology that is truly useful? Perhaps, maybe in medicine. Often, however, it is simply an advantage over an alternative life scenario (utility is something different, which does not need comparisons). If we look at the facts, when does new technology become important? From a business perspective, a crucial aspect is whether the technology can produce sufficient value: technology grows if there is a market. A good example is field-programmable gate arrays (FPGA) technology, a hardware that plays a role in scientific computing and industrial electronics: it was extremely useful because it allowed microprocessor production to be adapted to customer needs. Then came the graphics processing unit (GPU) technology, with ease and simplicity of use. Some still use FPGA, but it is not seen as essential technology; everyone, however, uses the GPU because, among other reasons, there are software tools (like CUDA) that have lowered the barrier to its use (regardless of the exponential growth seen today in generative AI). Business thus moves toward ease of use.

4.2.3.3 *Giuseppe Ragazzini: Machines of Creativity*

The third person we talked to about communication is Giuseppe Ragazzini, an artist who often experiments with the use of new technologies, reflecting on the difficulty of fully understanding their potential, and therefore communicating and realizing them. Giuseppe, in his reasoning, started from afar.

> We have been hunters and gatherers for about 70,000 years. Let's pause to think about what has happened in terms of technological evolution in the last hundred years, or even just the last twenty. We are facing the greatest change of all the preceding centuries: progress is not linear because it feeds on its own steps forward. In short, we have transitioned from the 'analog' man to the 'digital' man.

These two universes continuously coexist in Giuseppe's artistic work. An example is *Mixerpiece*, a creative application for teaching children the history of art that he commercialized in 2016. The application allows combining a range of elements drawn from great artistic masterpieces, offering creative possibilities. *Mixerpiece* uses a simple (but powerful) AI algorithm: when you shake the tablet, this combinatorial algorithm replaces the various pieces of the image based on the original scheme created by the user: the creative possibilities are virtually endless, making the application a real "creativity machine."

Mixerpiece also attests how quickly things are changing and how difficult it is to notice such change in our ordinary experiences. Giuseppe continues:

> In 2020, I gave a presentation in Florence where I addressed some themes related to the relationship between Art and Artificial Intelligence and recounted the birth and development of *Mixerpiece*. I remember starting my presentation by showing the now-famous portrait of Edmond de Belamy by the Parisian collective Obvious,

> from 2018; one of the first examples of art generated by a GAN (generative adversarial network), created by a system fed with a collection of about 15,000 portraits (dated between the 14th and 20th centuries). It was one of the first attempts to produce a work of art with machine learning, in its way impressive but still slightly pathetic, to be honest. In contrast, I showed a series of collages created by my app with its simple combinatorial algorithm, which seemed enormously more beautiful to me. Collage uses 'discrete' elements, separated from each other (and also pre-formed), while the difficulty in imitating painting lies precisely in the fact that it is 'continuous' and not made up of individual parts.

Only a few years after this event, it is impressive to think about the progress made by generative systems. But what is even more surprising is that most people are completely unaware: they are still entirely unaware of the scope and implications of this technology in everyday life. These generative systems today can process pictorial images not "simply" starting from "atomic" elements, as was still the case in 2018: what they produce is not a collage, but a fusion of styles, images, or elements into a perfectly coherent whole, now in many cases indistinguishable from the product of an artist. These are increasingly powerful and unpredictable systems; we could say that they demonstrate "emergent capacities," meaning they produce results irreducible to their basic components.

In his most recent experiments with AI, in which Giuseppe used an enormous dataset of "human" works to produce "beauty" through the new generative systems, the result is something new, unique, irreducible to the database; something "emergent," precisely. This makes one think (going beyond the artistic realm) that by feeding these increasingly powerful and fast machines with ever more vast datasets, we may witness something new, which our current mental categories do not allow us to conceive.

As an artist, Giuseppe has always praised chance, error, and imperfection as fundamental elements not only of the creative process but also of our evolution as human beings: if we were already perfect, then we would not need to evolve. But it is also true that in the artistic field in which Giuseppe operates, machines now perfectly imitate error and creative accidents and are capable of producing aesthetically stunning results.

At the same time, according to Giuseppe, we must recognize that often the mass aesthetics produced by machines are homogenized, and most of these works are immediately recognizable: after all, the public launch of these systems is still quite recent, and probably the aesthetic level of these works will improve more and more. We then return to a fundamental problem: the need to master technology, which in this case (that of AI) is much more complex to do. What will happen when anyone can create an aesthetically captivating work without creative effort and without knowing and experiencing the theme or subject? Some compare the phenomenon to that of photos that have been flooding us daily for years with the advent of cell phones. We are completely saturated by billions of photos (mostly ugly or insignificant), but we have not lost the value of a beautiful photo because what still counts is the human eye behind the lens. However, in the case of generative systems, the situation is different: the machines compensate for the photographer's eye by creating, if we wish, a photograph of high aesthetic quality inspired by the style of a great photographer—just know how to ask for it!, which is not always easy.

Speaking of art and technology, in recent years Giuseppe has often used the metaphor of a modern racing trimaran that flies on water, without contact, owing to two lateral aerodynamic wings called foils, the result of the most advanced technology: it is driven simply by the force of the wind and easily exceeds 100 km/h, but there is no trace of an engine; once it would have been a simple caravel or, at most, a fast brigantine. Giuseppe's hope has always been that art, evolving, continues to follow some of the "rules" on which it was founded: even if it makes use of increasingly advanced technologies, it should not forget its origins, just as the trimaran that flies on foils is still essentially a sailing vessel propelled by the wind. But does the metaphor continue to apply to new technologies like machine learning, or have we moved to a further level? After all, it is as if the trimaran on foils could now also generate the wind that propels it. What remains for us humans? Certainly, metaphorically, we still have the possibility to question the course to take, to maneuver this marvelous vessel (art) with more skill than others, and to arrive first at the destination, to win a regatta, to set a record. At some point, though, we probably won't even need to maneuver the vessel anymore: we will just have to follow a pre-established course or limit ourselves to establishing it.

According to Giuseppe, in a scenario where technology and AI reach levels of development that allow the creation of artworks and artifacts with extraordinary ease and precision, the value of the human-made artifact will take on a new dimension, becoming even more precious. This paradox is rooted in the concept of authenticity and uniqueness that only the human hand and mind can confer. When machines and algorithms are capable of producing objects and artworks in large quantities and of high quality, the human element, the error, the imperfection, the singularity of the creative touch will become an even more valuable distinguishing feature. Human creation brings with it not only the handmade and technical skill but also a history, an emotional and cultural context that cannot be replicated by technology.

In conclusion, Giuseppe continues, the real danger of this exponential evolution is that we are not able to culturally metabolize the progress we are witnessing. And what has been said here plausibly applies not only to the case of figurative arts but also to other forms of art or other creative processes. Our marvelous, ancient brain needs limits, the presence of obstacles, effort, error, and desire to function. Despite the great technological progress, our *Homo sapiens* brain is still programmed to reproduce, to seek food, shelter, and security, just as we did hundreds of thousands of years ago. We still don't know how to manage the consequences of our progress. We must educate ourselves and future generations to do so responsibly, developing both technological and human skills. Only in this way will we be able to face future challenges without putting our very existence at risk.

4.2.3.4 Leonardo De Cosmo: Journalism and Science Communication

After the conversation with Giuseppe, we finally thought we couldn't reflect on the communicability of science without discussing it with someone who deals with it from the media's perspective. We, therefore, turned to Leonardo De Cosmo, a science journalist for over a decade. Again, we present a condensed and edited version of our conversation.

According to Leonardo, there are two ways of understanding science communication. The first is based on the desire of the scientific world for science to be reported, on the one hand to educate the public and inform them on relevant ethical and political issues, and on the other hand to justify how public contributions are used. The second, entirely parallel, is that of the media, which sell content in the form of news and see science communication as a commercial tool (for TV, print, the Internet). The newspaper has to sell: it has no humanitarian vocation; it is run by actors who usually operate in a kind of news market. If the news does not interest readers, then that news has no value: it responds to the pure rules of communication within a specific economic horizon in which interest in a topic cannot be "forced" from above. Most science journalists try to reconcile these two ways of understanding their communication. But they are not always reconcilable.

There is also, according to Leonardo, the issue of relations with investors, which in turn relies on science communication. We can illustrate this with the case of graphene: the international scientific community worked on graphene for a long time, but investments in the sector also grew because the research was reported in a certain way.

In short, according to Leonardo, we can think of the media as predators of a sort, of varying sizes, that move in packs or alone. The news is their prey. A sector is preyed upon as long as there is something to eat: if the prey regenerates, fine, the predation will continue; otherwise, the topic will be abandoned. Continuing to prey on already known areas has advantages because the topic is already known and one knows how to move; if, on the other hand, nothing new and interesting is found, then it is better to leave the field in search of a new one. An example is the world of "quantum": when it was fresh, the news of "quantum supremacy" was a big and tasty meal, impactful, a technology straight out of science fiction incomprehensible to most but capable of tickling the imagination. Moreover, there was also a big company involved, already well-known. But will quantum be able to regenerate itself, find a way to continue providing news of real interest? If it does not, the media will change hunting grounds.

Another important element, according to Leonardo, is the factor of time. Reheated soup generally does not please anyone. But our memory is very limited, perhaps because we are overloaded or because we are distracted and truly ignorant: the fact is that after a few months or a few years, that same hunting ground that was abandoned can be revived, even with the same news already given, but in a renewed version. AI is very good at regenerating itself, also because the economic interests behind it are enormous, with large companies that, unlike researchers, know well the mechanisms of mass communication. AI enjoys "incremental" news: each time, it reveals a new piece.

Having established these points, we then moved on to talk about the "mission" of literacy in which science communication could engage. Leonardo argues that the educational messages we have managed to successfully convey in this area are very few and of poor quality. It is true that there are well-functioning channels and niches of communication, for example, on topics like chemistry or cosmetics; but, at the same time, these are also hotbeds of polarization. Usually, a sort of niche of enthusiasts is created, who more or less rigidly assimilate what is proposed to them, often with little critical sense (reaching, by the way, what we can hypothesize is the goal of the same channels). In short, if we compare the efforts to the results, literacy is almost a failure. Everything the media reports is forgotten within a few days:

> My experience has shown me that topics I have been writing about for years, even always the same over time, are actually forgotten, except perhaps during some peaks when, for whatever reason, they are much talked about. It is a phenomenon I observe in everything, even in the case of wars: a few years (if not months) pass, and we all forget the antecedents that perhaps had also received some attention not long before. Sometimes I talk to scientists from various fields and discover that they know only by pure chance that there is a space station orbiting the Earth, with people inside. This is something I might take as a given, especially since the station has existed for 20 years, and in one way or another, it is constantly being talked about. In this regard, for example, we journalists announce a cargo shuttle carrying supplies for the space station, taking it for granted that these are supplies for the people living there; or, we talk for weeks or months about an Italian astronaut who is on the station… Yet, even among people who work in science, such basic knowledge is often lacking.

Of course, not everything is like this: Leonardo points out that there is also a public that is open to new things and spans several science-themed topics, but it is probably a very small audience and very different from what we tend to imagine. We tend to greatly overestimate people's curiosity, desire to know, and explore, and this also applies to many scientists. In short, the educational impact of science communication, according to Leonardo, is very limited. Perhaps it is no coincidence, then, that the phenomenon of anti-scientism is emerging (a much more complex issue to justify, of course), given that communication is so difficult to convey.

Until a few decades ago, Leonardo continues, science was seen as something handed down from above, almost through authoritarian principles, infused by experts: those who received it had to take it as truth. The narrative of science spoke of a true revelation by scientific authority: a vertical model, linked to a wrong way of interpreting science. More recently, the idea has taken hold of wanting to tell what science is and how it works with a model that is no longer top-down, but more egalitarian and based on principles of open discussion, principles that, in some way, aim to convey the scientific method, made of uncertainties, changes, and many still unclear points. This approach to scientific communication is certainly more "correct" and democratic, but at the same time, it has created difficulties and increased the complexity of communication. Science is very complex: it actually has no linear path, no beginning, and no end, but is something that evolves, even on failed paths, in surprising ways. Science is a fascinating dynamic that does not fit with our mechanisms of understanding, with the cognitive schemes we need to follow a story. Would we go to the cinema to see a film with neither head nor tail? For all these reasons, telling a story when talking about science is difficult because the rules of creative writing and those of science are not the same and often diverge. (That said, Leonardo made sure to specify that he doesn't have a solution: he is only pointing out that there are various ways of interpreting what the right communication does and should do.)

The underlying idea, Leonardo continues, of these now-established models of science communication is for science to come out of its ivory tower and "go among the people," to involve and engage. This goal, however, risks causing real disasters because by aiming to disintermediate, it actually diminishes the professionalism of communication. Events like the European Researchers' Night, where researchers put on small shows, open labs, and engage children, are beautiful but also risky. It's one thing for a scientist to come down to the street, perhaps in front of a child audience, to do experiments once a year; it's another thing for someone to choose, as a profession, to communicate science and train for that purpose. On the one hand, the danger is that the scientist feels like a communication star and eliminates the emerging category of those who do science communication as a profession. Paradoxically, it can be counterproductive to push scientists at all costs to directly interact with the public; perhaps they are not good communicators, or they don't even have a basic understanding of the principles of communication, inclusivity, and so on.

Our conversation with Leonardo then ended with a sort of *praise of superficiality*, an apology for those who do science communication or journalism as a profession: those in this category have an overview that very few researchers have (because they tend to focus exclusively on their sector). It is, of necessity, an extremely superficial overview, unable to go in-depth, and suffers from all the mentioned flaws; yet, it manages to embrace more fields, and perhaps grasp affinities that someone engaged in going deep into a very specific topic cannot have. It's a bit like studying the Earth from space: clearly, from afar, you can't have the same details that can be observed *in situ*, but from up there, it becomes possible to understand global dynamics. According to Leonardo, the superficiality of those who do journalism is a competence that should have greater recognition, even institutional. There is much talk of the need for interdisciplinarity: science journalists have a lot of it. In recent years, entities and companies are increasingly limiting the presence of journalists, considering them a cost (thinking, for example, that they can still follow an event in streaming), but it is a mistake. Only by spending 24, 48 hours immersed with those working in a sector can criticalities and fundamental dynamics be grasped to then have a constructive overview.

4.2.4 Risk

Today, there are a growing number of startups founded by scientists in academic environments or within large companies. A crucial step in the growth of any startup is raising capital. Every type of investment carries a risk, which is based on the lack of information at the time the investment is made: the greater the uncertainty, the higher the risk (consider how many startups have a short lifespan or how market conditions can shift). Despite the uncertainty, investing in the early stages of a startup's creation can be enticing: there are growth opportunities, greater capital appreciation, and the ability to exert more influence on a still lean and malleable organization. Venture capital is historically the area of finance specialized in supporting high-risk, high-potential projects, investing money in early stage entrepreneurial ventures in exchange for equity. In 2023, global venture capital investment totaled approximately US dollars 345 billion, according to data from PitchBook and the National Venture Capital Association (NVCA). Within this global figure, the US market accounted for 170.6 billion in deal value (PitchBook & National Venture Capital Association [NVCA] 2024). This is quite different from bank credit or the rapid investment cycles typical of stock markets: venture capital has effectively supported many successful technology companies and has been one of the biggest (and possibly the most successful) drivers of innovation in recent decades. For this reason, it is important to discuss the topic here, considering risk as a key term to reflect on in making the scalability of scientific knowledge sustainable.

Venture capital is a very particular niche. We found it useful to discuss this topic with someone who has deep experience in the field: Raffaele Mauro, co-founder and General Partner of Primo Space, a venture capital fund that invests in innovative companies in the digital sector and the space economy.

According to Raffaele, over the past 30 years, the venture capital investment sector has experienced various cycles, with waves of enthusiasm often linked to specific emerging sectors such as e-commerce, "software as a service," the sharing economy, and green technologies. In times of expansion, risk tolerance and greater diversification among investors have progressively broadened the types of projects that can access financial resources. In this way, investment models for complex technologies based on scientific research have emerged and evolved, including robotics, space technology, AI, and new materials. These sectors previously had limited access to venture capital tools because they were perceived as too complex, too expensive, and with excessively long-time horizons.

Today, Raffaele continues, the time may be ripe to develop innovative investment strategies centered on the concept of "mathematical technology": companies and projects generated from the inexhaustible source of abstraction and generalization provided by mathematical thinking. In particular, we can consider mathematical technologies (math-tech) as the set of applications based on frontier research in pure and applied mathematics, physics, and theoretical computer science. For example, number theory is involved in cryptography, theoretical physics in quantum and nuclear technologies, and computational complexity theory in AI and new computational paradigms. Sectors affected by these developments include fintech (the digitization, disintermediation, and automation of finance), cybersecurity, and the use of big data and AI algorithms in commercial applications. Frontier research in these fields is often hybrid, with cross-pollination between academia and the private sector; there is usually a significant component of informal knowledge generated outside traditional boundaries. A significant example was the beginning of the cryptocurrency era, characterized by a substantial information asymmetry between a limited number of funds and individuals capable of understanding the technology in detail, going beyond slogans and marketing, and the larger set of actors willing to invest or operate in the sector. A different, more recent example concerns the world of AI, where, in addition to big tech companies that have been investing in this area for years, new hybrid organizations have contributed to evolving and spreading enormously successful applications.

In the future, according to Raffaele, we can imagine funds that invest specifically in mathematical technologies, explicitly targeting business ideas generated from both traditional research fields and innovative niches—from topology to category theory, from statistics and probability to graph theory. New types of institutions could develop in this sector, operating as both investment funds and research structures, focusing on recruiting the best minds in their respective fields and involving them in generating ideas and creating advanced technology transfer projects. Information asymmetries in frontier technologies could also be exploited by accessing the right human capital and adopting a specific combination of long-term investment horizons paired with the ability to generate short-term project results. Finally, it is possible to imagine a positive spiral in which technologies currently developed in the field of AI can accelerate the generation of results in the physical and mathematical domains, for example, through research areas like automatic theorem proving and its potential evolution toward increasingly creative construction of proofs and problem-solving pathways. We can thus hypothesize a venture capital model aimed at supporting projects and companies that translate the theoretical virtues of very abstract forms of knowledge into effective applications: as has happened multiple times in the last century, these forms of knowledge can generate insights and support technologies of enormous impact.

4.2.5 The Last Mile

Innovation and communication are both connected to another fundamental concept in technology, especially when it comes to deep tech: the last mile. Once we have developed a technology, how do we make it widely accessible? This is easy to say but extremely difficult to achieve, just like completing the last mile of a marathon. To define and better understand this concept, we sought the help of Fabio Veronese, a technologist and manager at Enel S.p.A., with 32 years of experience in technology and IT infrastructures.

A good starting point for discussing the last mile, according to Veronese, is to consider the state of the art in technology, particularly by asking: what should be changed in technology? Homer, in the eighth century BCE, sang of love, hate, power, and the thirst for knowledge. These impulses have not changed. The legendary Trojan horse was made of wood and was gigantic: now it has become a dematerialized and purely digital "horse." In short, the point is that our way of living, from Homer's time to the present, has evolved in a way unimaginable to his contemporaries, and the engine of this change has been technology. While technology has continually renewed itself over the centuries, however, human nature has fundamentally remained the same (which is why the exploits of Achilles, Hector, and Odysseus still resonate with us). Sure, there may have been changes in our perception of the world due to technology, but what makes us human has continued to rest on the same foundations throughout history.

When it comes to the growth of knowledge, we can identify various areas and phases of the scientific endeavor. One path is obviously to devote oneself to pure theory. Alternatively, we can focus on verification and experimentation. Or we can bring scientific results into business, for example, through startups or their use in companies, and then apply them on a large scale: those working in this latter area are not necessarily the ones who developed the theoretical part or created the startup. For example, the touchscreen entered commercial use through smartphone manufacturers, not through its inventors. In short, the benefits of a scientific idea or specific invention manifest when they can be delivered on a large scale. This is why the last mile is crucial: the part of the journey where technology becomes usable. There lies the benefit for humanity. With a "technofetishist" gaze, we see the enormous potential of technology that could benefit humanity but is applied poorly, or not at all, precisely due to a lack of proper management of the last mile phase; among these technologies, we might need to start counting renewable energy or the metaverse.

These considerations offer us insights into how to approach another issue we discussed with Veronese: what can be improved? First and foremost, basic research should be improved by investing more in universities and innovation (for example, also in startups trying to solve problems with a strong scientific component: investments do not necessarily have to be limited to universities). But where we risk losing the most value is in large companies: organizations that tend to have significant financial resources available and, therefore, could impact everyone's lives. The process of technology adoption in these areas is also quite mature, think, for example, of the adoption of evolutionary architectures (approaches to software development capable of evolving alongside technological changes and business priorities) or digital transformation. But the process is slow. We might also have a great idea, such as an adaptive control system for renewable energy: to turn it into a benefit for the community, we need to push and manage the transition from idea to user well, whether it's the company itself or society. Unfortunately, often those who generate technology do not know how to apply it, and those who manage it in large companies often lack technological awareness (or we might say maturity), fundamentally not seeing an advantage in the solutions that new technology can bring. The problem is precisely this lack of alignment in the innovation supply chain. Moreover, we often suffer from a bias (the so-called confirmation bias) that leads us to only talk to people who think like us. If we are among colleagues, we can build trust and manage bias; if a colleague asks you, "Have you heard about the new way to solve problem P?" then you have two possible answers: "I expected it" or "Great, I don't know it, explain it to me."

In companies that may have outsourced everything to maximize benefits, the market is regulated and, in some way, hierarchized; instead of exchanging information about how a technology works, we immediately talk about its impact with phrases like, "I'm bringing you this technology to improve your company's results by 20%." In this perspective, innovation is annoying and finds itself fighting against inertia of both decision-making and cultural nature.

There is, therefore, an underlying hypocrisy. Technology disrupts already established mechanisms and is, therefore, not well-received because it is not easily assimilated. Yet, few people are against mature technology, although it is clear that skeptics exist and will always exist. Practically, once again, at the level of large companies, where there is the opportunity to apply technology (due to financial and operational capabilities), there is rarely openness—and here, it should be noted, it is not a question of administrative and managerial capabilities in managing change processes. There is fear because technology could change and reach a new optimal point that may be perceived as better for the system as a whole, but not necessarily for the specific company. In the famous board game Risk, the rule is that in the event of a tie between the attacker and the defender of a territory, the defender wins. If a player wants to attack, they need more tanks. In conditions of equality, those who want to resist are favored. To put it humorously, this is either an analogy or an exaggeration of Brandolini's law: the effort required to refute false information is greater than the effort required to produce it. Returning to companies, those who want to oppose the adoption of technology simply need to use inertia: they don't have to do anything. It's a kind of second law of dynamics applied to the business world. With this metaphor in mind, let's imagine the mass of large companies. If set in motion, they can have a formidable momentum, but getting them moving requires enormous effort, which is not always possible to achieve. In conclusion, another aspect that could improve the situation might (perhaps) involve more strongly adopting emotional communication strategies aimed at making technology available—telling the technological opportunity with expressiveness and effectiveness so that it resonates with decision-makers—or embracing agility principles. Much has been written about this topic, of course, as it is central in discussions about corporate strategies to respond to market changes or new and emerging opportunities. That said, it is not obvious how much and to what depth the same topic is addressed in the context of technology development itself.

A different discussion, moving further upstream in the knowledge chain, concerns so-called high-impact technologies: simply put, these are technologies whose sudden absence would have negative consequences on people's lives. Veronese pointed out the concentration of capabilities associated with certain new technologies and how, at times, innovation ends up being blocked before it even reaches the last mile, not to mention that paradoxically, the block is created by the very dynamics that should create value. The accumulation of significant capacity (or, to use a strong term, power), history demonstrates, leads to an almost irresistible temptation to misapply it, seemingly as a natural, albeit often unconscious, impulse. A common illustration of ongoing discussions involves the undefined regulatory frameworks for modern tech. And recently, it has become clear that to train new foundational models, for example, requires energy and hardware on such a scale that only a few organizations can afford them. This concentration mechanism of some high-impact technologies risks becoming a brake on innovation and can lead a company to become a victim of its own success.

For social networks, we know that the benefit they create is proportional to the number of users (this is an empirical fact: the utility of a service increases with the number of users); according to network economics, on the other hand, each participant increases the platform's value.

In the case of social platforms, to give a concrete and fairly simple example, a drift may occur where, once a huge number of users is reached, the platform's managers begin to think that their service is no longer an option but an obligation: the platform is "closed," and at this point, users end up paying a subscription for specific services or lose benefits such as the ability to manage their online time without interruptions and distractions (see Zuboff 2019 for an in-depth treatment of this topic). On the other hand, those who manage the service can take care of important aspects ranging from cybersecurity to management aspects to coordinating the development of new features. Veronese believes in the value of protocols—the procedures, using the broadest possible term, used by machines to communicate and process information:

> In a world based on protocols, everyone can do everything. Protocols are open, so I am free to 'do' but cannot 'prevent.' This is the fundamental advantage of systems based on protocols. I am referring to computing, but it is a broad concept.

For example, during the early days of the Internet, the Internet Relay Chat (IRC) protocol allowed real-time message exchange (instant messaging). IRC is an open protocol: it is publicly available, and does not belong to any particular organization. The fundamental benefit of an open protocol is that users can use it at no cost and modify it as they see fit.

Touching on these points, we are now tempted to consider an analogy with open protocols and try to apply it to science. To do so, we choose the extreme case of total decentralization. What would the result be? To try to answer, it is worth introducing (perhaps a bit forcefully, at least seemingly) the role that so-called citizen science can play in creating models of science and innovation that include the perspectives of multiple audiences. To present the topic, we discussed it with Muki Haklay, a professor of geographic information science at University College London and an expert in citizen science and participatory research.

We first asked him to define citizen science, which can be characterized as the participation of non-professional scientists in research projects. The non-professional figure is important because it has been shown that many participants in citizen science projects have a terminal degree (e.g., PhD), but they are not professionals in the field of the project they are participating in, or they do not work within a research institution.

We then moved on to discuss the goals of citizen science, which vary widely along a spectrum. At one end, it can provide scientists with resources that would otherwise be impossible to obtain, which is closely related to the practice of crowdsourcing. Imagine, for example, not having enough computing resources: we can collectively pool them through voluntary processing by many participants. Or imagine needing observers everywhere, as happens when collecting meteorological data. At the other end of the spectrum are cases where people come with their research questions, such as a community facing environmental justice issues. All of this is part of the same picture. The vast majority of projects are led by scientists, while do-it-yourself and community-led research are, in most cases, small-scale projects.

Citizen science, Muki continues, promotes open science. Now, when scientists think about open science, they usually think of open access, open data, and open-source software. However, the ultimate openness of science is not just about the club of scientists; citizen science is probably one of the best ways to think about open science: a science done by a broad public that innovates through a high degree of engagement and participation.

We then asked Muki to give us an example of an ambitious citizen science project; he cited a case from the UK concerning birds: the Big Garden

Birdwatch. At its peak, the project involved a million people, about 1.8% of the UK population. It is the largest wildlife survey in the UK's gardens, where quantity turns into quality. It has been active since 1979, and about 700,000 people participate yearly. Over all these years, it has helped and continues to help us account for the prevalence of different bird species and populations across the UK.

Moving on to the future of citizen science, Muki believes that in highly educated societies (as is the case in the European Union), it has been shown that many more people are potentially willing to participate in citizen science than currently do. Another important aspect is the importance of inclusivity. The opportunity for marginalized groups to be involved in research is important: Muki has been involved in work that has shown that even people with very little technical literacy can actively engage in the co-design of citizen science. Another interesting aspect today and potentially in the future is the interaction between people and technology. Think, for example, of a PhD project where the goal is to classify galaxies: after viewing 50,000 images of galaxies, even the most passionate astrophysicist runs out of energy; with the help of citizen science, however, larger and more reliable training sets and data sets can be obtained. Here is another possible role for citizen science, not to mention that it could also help understand when to get help from machines and when it is preferable to turn to humans. Even the best computational methods today require people to help improve systems.

Finally, as for the main challenges facing citizen science, according to Muki, the biggest one is acceptance by professional researchers; currently, the level is still low. There is also the issue of financial support: today, there is relatively little funding for citizen science, while reaching marginalized groups and people with local and indigenous knowledge is crucial. In this sense, we should think of science as the least bad method we have for accumulating human knowledge in every field—for example, about medicinal herbs traditionally and successfully used by local communities. Science requires respect for all forms of human knowledge, and citizen science offers a space where this can happen. In this way, the challenge of integrating citizen science with the rest of scientific research has the potential to transform knowledge as a whole.

4.2.6 Ordinariness

We close this section with a reflection on how machines could be integrated to simplify and transform everyday experiences. We discussed this with Paolo Dello Vicario, founder and CEO of ByTek, an Italian martech company (which develops technologies to optimize marketing processes) that deals with applying AI to performance marketing.

In recent years, Paolo explains, technology has enabled increasingly simple and quick shopping and consumption experiences; we have witnessed a gradual simplification of interfaces and an increasing spread of personalization systems that tend to anticipate a user's next move, reducing the cognitive load required to complete actions that, until recently, presented themselves to us in all their complexity.

A purchase, finding a movie, staying up-to-date with the most relevant news in our sector—these are all activities that require less and less effort, meeting the paradigm summarized in the title of a book that is a milestone in the usability of digital products: *Don't Make Me Think.*

In 2000, Steve Krug already anticipated the principles of digital design, which increasingly pushed this concept to the extreme, simplifying the lives of billions of people with the myriad systems and technologies that today are part of our daily lives.

This gradual simplification finds its culmination in the alignment of the language of technology with so-called natural language, the one spoken by humans: barriers are eliminated, interfaces are reduced to the bare minimum, and a new phase of technological development begins.

With the widespread adoption of LLMs and generative AI in companies, it is no coincidence that the race to simplify interfaces with the end customer is even more ambitious. The major digital companies are constantly questioning how to rethink their interaction with users; they are introducing personalized interaction systems in conversational modes that greatly accelerate that process of delegation of choice that we have been witnessing for years; and when ideas of true personal assistants (or, one might say, digital concierges) appear on the market, the step toward complete delegation not only of actions but also of purposes becomes very short. These concierges, based on our past interactions and a set of other information, propose not only to "search for us" but also to suggest programs and actions and perhaps execute those actions themselves, interacting with that external world that now has no linguistic barriers for machines—languages, after all, are aligned.

Without thinking of apocalyptic and surreal scenarios, when it becomes possible to delegate the organization of a vacation to an automated system that proposes a travel idea among the infinite possibilities given by the combination of itineraries and unexplored places, perhaps going from ideation to execution in a matter of seconds, we are actually opening the door to an entirely new world.

Will we live increasingly guided experiences directed by these systems, which, in a self-feeding mechanism, will lead to a flattening of tastes and purposes, or will we have more time to devote to defining those purposes? Are we delegating too-important activities to the machine, or are we building a mechanism to find balance in that paradox of choice that Barry Schwartz theorized back in 2007?

Certainly, in this world, companies that want to be competitive will increasingly need to invest in their ability to be an appealing, or at least possible, choice for the technological concierge. An increasing share of companies' revenues will, in fact, be attributable to so-called machine customers, autonomous in managing transactions on behalf of someone else.

More and more choices will thus be delegated to machines, which will themselves become, in a sense, the customers of companies; customers on whom to make investments, allocate budgets, and direct the development of their technologies. These machines will need to be increasingly efficient to satisfy other machines and ultimately move money and investments.

If it is true that science is driven by technological developments, that investment choices in technology and science are ultimately guided by market demands, and that these market demands will increasingly be determined by technology itself, then we are facing an important and highly significant paradigm shift.

Technology, which could serve as a muse accompanying and supporting *Homo sapiens* in his activities without apparently influencing the definition of purposes, becomes an invisible hand guiding him through the market.

The debate to ensure that a wonderful moment of technological development can also be, this time, a useful support for the world is open. Perhaps we will discover that not everything has to be reduced to a simplification of experience, but that appreciating and embracing the complexity of this world is part of the experience itself.

CHAPTER 5

Epilogue

Notes on How to Guide the Construction of Machines for Future Science

The ideas in this volume emerge primarily from the daily practices of doing science and building technology, in dialogue with colleagues who engage in these fields. In this closing and short chapter, we reflect on how to apply what we have discussed to guide the construction of machines for future science.

A key insight from Chapter 4 is that it seems almost inevitable that science is moving toward development on a hyperscale. This raises the urgent need to rethink the perspective of science done in collaboration with machines, where these machines are not merely seen as laboratory tools or calculators but as assistants possessing technical capabilities that surpass our own. For centuries, we have been accustomed to thinking that there is a close connection between scientific knowledge and machines—whether because scientific knowledge relies on machines across various branches and fields, from telescopes to microscopes, from air pumps to calculators. However, a different possibility is now emerging: that some branches of science may be carried forward by machines with relative autonomy. This is a far more radical scenario than what we have been accustomed to, which has led colleagues to look with uncertainty at its possible developments (see Nielsen 2025).

DOI: 10.1201/9781003676614-6

This prospect is more than a mere technical evolution; it represents a fundamental challenge to our very conception of the scientific endeavor. It forces us to ask profound questions about the nature of discovery and knowledge evolution in the context of science. What does it mean for a scientist to "understand" a phenomenon when valuable insights are generated by a non-human intelligence? Where does the line between a tool and a collaborator truly lie? Yet, this disruption also contains immense promise. By delegating the labor of computation and data analysis, it may free human scientists to focus on what they do best: exercising creative intuition, asking novel questions, and applying ethical judgment. How we navigate this redefinition of the researcher's role is perhaps the central metascientific task of our time. Of course, this task must be approached with caution, consciously avoiding the traps of technological hype and distractions like the anthropomorphization of machines. The challenge is a practical one: to pragmatically identify how to use these novel technologies fruitfully. This requires a deliberate effort of integration and learning, just as it once took work to discover how the pocket calculator could best be incorporated into the scientific workflow.

We should be invested in thinking about this development *not* (or at least not only) because it improves technologies focused on creating services that bring immediate benefits while still requiring research and development, *nor* because it regards speculative technologies that are predicted as the consequence of engineering challenges yet to be overcome and scientific discoveries yet to be made. To be developed on hyperscale are technologies at the heart of various startups straddling academia and the business world, driven by the scientific talent of their founders, though often with vague and difficult-to-visualize business plans.

Stepping back to the theme of Chapter 3, we find another important indication: the need to renew the way science is conducted, particularly by rethinking the paradigm of the scientific article and renewing research funding methods.

Is there a good way to combine these two needs, renewing how science is done through a renewed relationship between researchers and machines?

5.1 MACHINES AND PEOPLE

Jesse Han, CEO of Morph Labs and a former researcher at OpenAI, mathematician, and expert in automated reasoning and artificial intelligence (AI), gave us a compelling image from which to start writing this final part of the volume. Jesse believes that the future of science will closely resemble the future of software. We can imagine that the universe is performing a very long computation, which we can somehow observe. According to Jesse, software fundamentally involves designing interfaces for systems with complex computational semantics: the various parts of the system interact in ways we cannot predict but that have meaning. Science, as the study of reality, has much in common with how people create software. At the point where we are and where we are headed, Jesse suggests, the practice of science will be the application of intelligence to the search for truth. This is precisely the type of activity that occurs when creating software. For example, finding the right level of abstraction to express an idea in code is much like the process of searching for the right set of abstractions to explain a natural phenomenon. Concerning the organization of scientific information and interaction among scientists, we will see an ever-increasing wave of specialized and general AI performing increasingly complex tasks under human scientists' supervision. The final state of systematic scientific investigation is similar to the final state of mathematics, which is the same as the final state of software creation. As increasing amounts of computing power concentrate in individuals' hands, a single scientist will be able to supervise a system on a planetary scale, like a (potentially distributed) computer with millions of intelligent subprocesses, all seeking the right abstraction to describe the universe.

The image of scientific research suggested by Jesse is in stark contrast to the one many of us (perhaps almost all) have. When we talk about science, we usually think of a series of activities carried out by people: fallible, thinking beings driven by emotions, "emotional machines that think," to use an expression from neurobiologist Antonio Damasio (1994); beings necessarily vague and distracted, who sometimes end up maximizing different and accessory functions, such as profit, visibility, prestige, or some other personal interest, while at other times they proceed with their work by inertia, forced to conform to the uncontrollable dynamics of academia, the corporate world, or simply driven by the flow of what society prescribes.

The scenario of a science conducted, at least in part, by machines raises obvious concerns about its effectiveness and, ultimately, sustainability. As we saw in Chapter 4, it is important to understand this term broadly, as a *theory of resources*. This approach allows us to discuss in various aspects how science can be optimized, i.e., improved in relation to some parameter, through the use of machines. To do so, we must also account for the sustainability of the scientific enterprise's core *epistemic* and *social* values.

The risk—in Han's vision—is not just inefficiency, but that a hyper-optimized, machine-driven system could simply amplify the very human flaws we seek to overcome—entrenching bias from its training data, mistaking metrics for truth, or pursuing intellectually sterile paths with formidable speed. An unsustainable system is one that erodes trust, discourages intellectual diversity, or prioritizes cheap outputs over impact, no matter how resource-efficient it appears to be.

But how could we today (re)configure the relationship between machines and scientists in a sustainable way?

5.2 SCIENCE FOR A HOUSE OF SOLOMON

We don't have a magic wand to solve the question just raised. The best we've found is to put the problem in perspective through a four-century-old lens used by Francis Bacon in *The New Atlantis*. Bacon's incomplete work provides an ideal backdrop for rethinking the relationship between science and machines for contemporary science. This small volume, considered today a classic of scientific thought, was published by William Rawley in July 1626, shortly after the author's death. In the text, Bacon tells of a fantastic House of Solomon, a utopian institution that diligently advances knowledge and invents new technologies, working in the interest of society as a whole. And indeed, a few decades ago, Paolo Rossi wondered whether contemporary science resembled the House of Agrippa or the House of Cardano more than the House of Solomon:

> The 'natural philosophers' of our time have had to work, in many cases, in environments quite different from that of the House of Solomon. A research founded on secrecy rather than collaboration and the publicizing of results, aimed at the power of a state or a social group rather than the well-being of the entire human race, a technique reduced to thaumaturgy or an instrumental aid to a superstitious worldview, an exploitation of nature founded on the exploitation of man: all this is undoubtedly closer to the ideals of Agrippa and Cardano than to those of Bacon and Galileo.
>
> *(1962: 23, our translation)*

Today, more than 60 years after the first edition of Rossi's text, how should we interpret the intertwining of science and technology? And how can we ensure it is sustainable and thus more like an emanation of the House of Solomon than that of Agrippa or Cardano?

5.3 DRAFT OF A SOLOMON'S ALMANAC

Taking the questions just posed literally, we can begin a small "literary game." Let's imagine ourselves as two employees at a House of Solomon for contemporary science, tasked with drafting an almanac that succinctly and directly summarizes the relationship between science and machines for science—*Solomon's Almanac*, if you will. If we were responsible for such an Almanac today, then what would we write in it?

In ancient Arabic culture, the term al-manākh referred to a type of astronomical table used to track the positions of the planets. Later, almanacs became texts that provided essential information on a certain topic. Now, without delving further into these fascinating historical details, we can think of our Almanac as a draft meant to give shape to metascientific discussions on the relationship between science and technology, a draft we will try to outline by summarizing the most significant points discussed in this volume.

What Is Science? Following some intuitive ideas, we can think of science as an activity tasked with describing and predicting phenomena through the most compact expressions possible and results that are verifiable and communicable by both people and machines. The importance of language in carrying out this activity cannot be overstated: a scientific result is a proposition, expressed in a formal language, that has the properties of being mathematically true or experimentally verifiable. This makes a scientific result a transmissible element of knowledge through expressions of a language, an element for which, until its discovery, we had neither proof nor a reasonable psychological guarantee.

(Brief addendum. Our definition of science is debatable: whether its focus, particularly without an explicit demand for "explanation," weakens the scientific endeavor. Critics may argue that true science aims for a deeper understanding of why phenomena occur, not just what will happen or what is the case. For example, they may contend that opaque "black box" AI models, while excellent at prediction, fall short of providing the causal insights necessary for actionable interventions and robust theory building, potentially rendering scientific findings less satisfying and less "scientific" in a traditional sense.

However, a counter-argument suggests that this emphasis is not a weakening but a pragmatic re-evaluation of science's goals in the age of complex systems and advanced AI. For someone, particularly those aligned with constructive empiricism or instrumentalism, empirical adequacy—the ability to accurately describe and predict observable phenomena—is sufficient. The argument here is that a complete causal explanation, in general, might be impossible, and that accurate prediction, even without full transparency into "why," holds practical value. From this perspective, explanatory theories can be seen as "useful fictions," and the ultimate goal of science might be control over natural phenomena, for which effective prediction is paramount.

Science Done by Machines Today, we are witnessing a transition in our interaction with machines, which at this point must be seen as an integral part of the metascientific discourse. Machines now go far beyond calculators: they are tools for searching and cataloging information, but also for obtaining and verifying scientific results autonomously or under human guidance. It would be useful if our interaction with machines could comfortably take place in natural language, but for now, this remains far from being not only achievable but also even conceivable in detail.

What Is Technology? Technology serves as a bridge between the thinkable and the buildable world, suggesting construction procedures that solve a problem and often subvert what we might call a natural state. For example, technologies extend life through medicines; they allow us, through a car, to go uphill against the force of gravity; they enable us to move at speeds impossible to reach with just our legs; they make us fly and convert solar energy into electricity. Sometimes they even paradoxically bring us closer to what we think of as the natural state, distancing us from an *artificial natural state*, to use a perhaps imperfect term, created by technology itself. Just to cite one voice among many, Peter Lunenfeld, a professor at UCLA, even sees a real vicious circle in technology: "We will use technology to solve problems created by the use of technology, but the solutions will lead to new problems" (Pew Research Center 2020).

Science and Technology I Technology is not science. This is a banal point in some respects but important. Technology is instead a form of knowledge that provides procedures for building machines or operating them. The language of technology is, consequently, that which is needed to describe the procedures necessary for it. Among technologies, some we can call "scientific" because they are based on scientific knowledge (they partially use formal language) and are tested through the experimental method.

Science and Technology II Technology does not aim to produce truth but rather a guarantee of the belief that a certain procedure will lead to creating and operating a machine that reliably attempts to optimize a certain function.

The Value of Technology In this sense, unlike science, the value of technology is only partially epistemic. For example, a therapy against a type of cancer does not teach us something new about cancer but serves to save many patients' lives. More generally, for every scientific proposition, there are many, potentially infinite, scientific technologies. The value of technology, therefore, lies elsewhere: in the value of the procedure; in the value of the machine; in the contextual value of its function (which, in turn, depends significantly on the narrative), what we have called the "utility function" of a technology.

The Role of Machines Traditionally, scientific knowledge enables technological development, and technology contributes to advancing science by building machines that assist research in its various stages: experiment design, execution, analysis, calculations, formal procedure, and result verification. Today, we are entering a phase where certain machines can directly guide scientific research, for example, through literature analysis, organization, hypothesis formulation, and verification; it becomes plausible to think that machines could contribute to refining scientific methods themselves in ways and with consequences that are still exploratory.

The Role of Metascience The study of doing science is a recently formed field of scientific inquiry that we can call metascience, aimed at systematically studying how science evolves to predict how it will evolve and correct its course, if and how possible; one could also refer to metaresearch, which focuses on how research is conducted.

The Scientific Article The scientific article is the main tool for regulating not only content but also social aspects of doing science, including competitions, career advancements, and funding. However, the system of publishing articles has significant flaws: the most evident are the evaluation times and methods and the publication costs, but there are even greater problems, such as directing research toward areas of little impact or distributing information only partially useful for the development of science itself. At least in some sectors, the scientific article should be replaced by more compact, traceable, and therefore, reliable forms of writing and communication. As we rethink these formal structures, other communication duties become even more vital. Chief among them is outreach work, which remains an almost obligatory responsibility for scientists to ensure that science and technology are woven into the global collective culture.

Science for Science's Sake This principle asserts that the pursuit of knowledge is a valuable end in itself, independent of any immediate application.

While it may seem like a luxury in a world driven by practical demands, it is the bedrock of long-term progress. History shows that the most transformative technologies often spring from discoveries made out of pure curiosity, with no practical goal in sight. Michael Faraday's foundational research on electromagnetism had no foreseeable use in the nineteenth century, yet it became the basis for the entire modern electrical age. Protecting this space for unfettered, curiosity-driven inquiry is not a rejection of utility, but the most profound investment in the unforeseen utilities of the future. Furthermore, the value of pure science transcends technology. It fundamentally alters our understanding of the world and our place within it, generating new ways of thinking that become part of our collective cultural heritage. In a research ecosystem that increasingly demands immediate, measurable impact, vigorously defending the principle of "science for science's sake" is essential. It is a commitment to the idea that the expansion of human knowledge is a primary good, one whose full benefits—practical, intellectual, and cultural—can never be fully predicted, only nurtured.

Science in a Relational Structure (Science beyond Scientific Literature) To truly scale in a sustainable way, scientific knowledge must be liberated from the linear, narrative format of the traditional article. We must begin to imagine its future as a formalized, computable structure—a dynamic system that represents not just discoveries, but the intricate web of relationships between them, workflows, data, and even mistakes.

This system could take many forms: a knowledge graph where ideas are nodes and logical dependencies are edges is an intuitive starting point, but it could also be a verifiable logical framework built with tools like Lean, or a semantic web of machine-readable assertions. Whatever its final form, the primary goal of such a structure is to make the architecture of science explicit and machine-readable. This would enable a new class of automated tools to navigate, analyze, and even extend our knowledge. Machines could autonomously verify the logical integrity of new claims, identify hidden connections between distant fields, reveal the true intellectual foundations of a theory, and even suggest novel hypotheses by identifying gaps or unexplored adjacencies in the structure. This would allow us to develop far more meaningful metrics of impact, moving beyond simple citation counts to measure a discovery's "structural importance" or its role in opening up new territories of knowledge. This does not render the human-readable article obsolete. On the contrary, it refines its purpose. The traditional article (or its future equivalent, such as a blog or review) would become the essential space for the elements that resist formalization: the narrative context, the historical background, the creative leaps of intuition, the lessons from failed experiments, and the speculative framing of future research. This creates a powerful hybrid ecosystem—a robust, verifiable, and machine-navigable core of formal knowledge, enriched and humanized by a surrounding layer of narrative and interpretation.

The Breakthrough What counts as a breakthrough, and how do we encourage it? First, we need a broad view in defining this concept, including breakthroughs in scientific methodology, knowledge organization, and techniques. As Terence Tao reminds us:

> good mathematics is not merely measured by one or more of the "local" qualities listed previously (though these are certainly important and worth pursuing and debating), but also depends on the more "global" question of how it fits in with other pieces of good mathematics, either by building upon earlier achievements or encouraging the development of future breakthroughs.
>
> *(2007: 632)*

Furthermore, greater efforts should be directed toward encouraging breakthroughs, promoting education and evaluation methods that value not only the individual qualities of a scientific product but also their "global" placement, as suggested by the above passage.

Funding Processes In funding processes, a parallel problem to that of scientific articles arises: evaluation is too costly for both those evaluated and those who evaluate, with the problem that the latter often fail to express a reliable judgment. Funding should be more focused on encouraging breakthroughs. Funding must explore new methods that can include private capital more substantially. Hence, the push toward forms of funding and investment for commercial purposes is less concerned with short-term profit, almost a hybrid between foundations and startups.

The Context Knowledge cannot be associated with data but must be associated with people. Thus, even in scenarios where the use and centrality of machines in science become routine, the human component remains essential and irreplaceable, barring major innovations beyond the visible horizon.

Scalability The scientific enterprise has always scaled its communication, from the printing press to digital archives. The new frontier, however, is to scale the core cognitive acts of discovery itself. This means using new technologies to scale rigor through automated proofs, creativity through AI-driven hypothesis generation, and experimentation through robotic labs and high-fidelity simulations. The ultimate aim is not just to accelerate science, but to build the cognitive infrastructure needed to attack problems of a complexity that was once unimaginable.

A note is due. The call to "accelerate scientific discovery" has become a modern slogan, but it is a phrase that must be handled with care. We need to distinguish this goal from a form of technological *accelerationism*—a blind pursuit of speed for its own sake. The primary aim is not simply to do the same science faster, but to equip humanity to tackle problems of a fundamentally higher order of complexity. The second, complementary goal is to radically speed up what Thomas Kuhn famously identified as the daily work of "normal science": *puzzle-solving*. This is a kind of parallelization. As Kuhn noted in *The Structure of Scientific Revolutions*, "Under normal conditions the research scientist is not an innovator but a solver of puzzles…." This is where the vast majority of scientific labor is spent. It's a domain where machines can offer an advantage: the painstaking work of testing hypotheses. By partly automating the puzzle-solving machinery of science, we give more space to the human intellect to focus on asking the right questions.

Words to Bet On Scalability involves risks that are not always predictable. One way to address it and face these risks is to equip ourselves with keywords to guide the transition toward a new phase of collaboration between machines and scientists for advancing scientific knowledge. Only in this way can the scalability of knowledge research perhaps go hand in hand with the sustainability of this process.

As a new phase of collaboration between machines and scientists takes shape, the exact nature of these advanced partnerships remains largely undefined. While the *concept of coalitions of agents*—where human and AI intelligences work in concert—offers a compelling vision for scaling scientific knowledge, the precise mechanisms and technologies that will underpin these coalitions are still emerging. One promising direction involves leveraging language models as sophisticated statistical heuristics. However, recognizing the inherent statistical nature of language models, formal verification emerges as a critical, complementary discipline. By employing rigorous methods from logic and mathematics, formal verification can meticulously check the consistency, correctness, and validity of the insights and conclusions drawn from these models. This dual approach—combining the exploratory/heuristic power of AI with stringent validation—could offer a pathway to managing the unpredictable risks of scalable knowledge discovery. Yet, it's also vital to remain open to the possibility that the ultimate solutions may involve something entirely different, breakthroughs we haven't even conceived of yet, highlighting the dynamic and evolving nature of this field.

5.4 A CALL TO ACTION

The development of scientific research is a complex process involving many very different actors, from research groups and communities to the institutions that coordinate them, through the actions of individuals and citizens. This draft of *Solomon's Almanac*, once reviewed and integrated based on specific needs, aims to be a document for rethinking and negotiating the interplay between science and technology, with a core belief: that the future of science must be a collective endeavor, shaped by all sectors of society, and that its knowledge must, in principle, remain universally and effectively accessible.

While it is sometimes important to focus on how scientific innovation happens or could happen in certain countries, for example, those that lead it or those where it could have the greatest impact, it is crucial to think about the topic on a planetary scale. The future of research concerns global civil society, and therefore the guiding principles for scaling must be discussed on this same scale. Here we agree with the ideas proposed by Nielsen and Qiu (2022) and by Nielsen himself in other publications and written and oral testimonies: metascience is fundamentally compatible with a vision of science open to all of society, which cannot be restricted to the interests of specific groups, companies, or countries.

A corollary of a commitment to open science is that the conversation about its guiding principles should involve all sectors of society. To translate the ideas, we have made explicit in the text into actions, we will end the volume with a direct appeal addressed to different groups of stakeholders in scientific innovation.

For Governments and Institutions: Scientific research emerges from our investigation as one of the main vehicles of innovation for human societies. Science is an excellent investment for anyone, including governments, enterprises, and private individuals, and serves as a fertile ground for innovation. An investment, moreover, is particularly fruitful when it establishes a virtuous cycle that leads science to influence technology and the latter, in turn, to accelerate research. (And, mind you, we are not saying that science has never played this role at other times in its history; indeed, it may be a distinctive feature, if not inherently necessary, of much scientific research.) But how can the interplay between science and technology be made scalable from a government's point of view? What rules favor the scalability of innovation without discouraging investment in research and technology and without stifling the work of scientists?

A multi-pronged strategy, drawing on insights from organizations like the Organisation for Economic Co-operation and Development, is essential. The foundation of this strategy must include data and computation. Governments and funding agencies have a critical role in supporting the creation of high-quality, large-scale public datasets managed according to FAIR (Findable, Accessible, Interoperable, Reusable) principles (see Section 1.3).

Concurrently, policy must address the prohibitive cost of computation by ensuring academic researchers have affordable access to supercomputing resources. Beyond infrastructure, the very organization of collaboration must be re-engineered. The most pressing scientific challenges are often interdisciplinary, yet funding structures frequently reinforce the silos that separate fields. To counter this, funding must be directed toward broad, multidisciplinary programs. The NSF's Convergence Accelerator program, which fosters such integrated teams, serves as a valuable model.

To ensure that new knowledge translates into meaningful real-world benefits, the research community must become more systematic about how it scales innovations. The "voltage drop"—where an intervention's impact diminishes significantly when moved from a controlled trial to a messy, real-world context—is a common and costly failure. This necessitates actively funding and promoting the "Science of Scaling" as a formal field of inquiry. It means moving beyond discovery and applied science to formally recognize and support scaling science as a distinct stage in the innovation lifecycle. Such a field would be dedicated to answering crucial questions: Under what conditions should an innovation be scaled? How can we define an *optimal scale* that balances magnitude and sustainability? And how do we coordinate the complex web of actors needed to achieve impact? Investing in this science of implementation is essential to increase the societal return on our vast investment in research.

These individual strategies, while necessary, all point to a deeper, overarching challenge: the *meta-problem of governance*. The scientific enterprise is a complex adaptive system, and optimizing it requires a coordinated, system-level approach. Implementing one recommendation in isolation—for example, changing incentives to reward risk-taking without also providing the compute and data infrastructure to make those risks tractable—is unlikely to be effective. The ultimate challenge is to design and manage the entire research pipeline as an integrated whole, creating a new function within science policy explicitly tasked with this strategic oversight. This role would be responsible for pacing acceleration and scaling, identifying system-wide bottlenecks, and ensuring that the powerful new engines of discovery are steered toward destinations that align with long-term societal values.

For Society: Scientific research is fruitful when it establishes a virtuous relationship with the needs of the societies from which it emerges and in which it operates—a point that is not necessarily self-evident. This consideration is well expressed by Daniel Sarewitz in his 2016 essay "Saving Science," published in *The New Atlantis*. Sarewitz argues that the only way to ensure a future for scientific research is to ensure that it remains firmly connected to the needs of the human societies that use it. Specifically, according to Sarewitz:

> First, scientific knowledge advances most rapidly, and is of most value to society, not when its course is determined by the "free play of free intellects" but when it is steered to solve problems — especially those related to technological innovation. Second, when science is not steered to solve such problems, it tends to go off half-cocked in ways that can be highly detrimental to science itself. Third — and this is the hardest and scariest lesson — science will be made more reliable and more valuable for society today not by being protected from societal influences but instead by being brought, carefully and appropriately, into a direct, open, and intimate relationship with those influences.
>
> *(2016)*

This is why, to understand the state of scientific research today, we considered it essential to report the voices of colleagues who participate in various ways in scientific production and who mediate with different needs and sectors of society. For research to be truly scalable, it must be rooted in a constant and renewing dialogue with the social fabric.

Central to fostering this virtuous relationship between science and society is the cultivation of a scientifically literate population. A deeper public understanding of scientific processes, their limitations, and their immense potential is vital for informed decision-making and for building sustained trust in research. Educational institutions, media, and scientists themselves have a shared responsibility to demystify scientific inquiry, communicate findings transparently, and engage citizens in a meaningful dialogue about the role of science in their lives and the collective future.

For Scientists: If we have written this text, ultimately, it is because we believe there is still room to manage, in a metascientifically virtuous way, the relationship between science and technological innovation. We have tried to outline good conduct for setting up such management. We did so primarily for ourselves, out of a metascientific need that we believe should guide anyone interested in science. What paths forward emerge from this analysis? They are not firm directives, but rather guiding questions intended to provoke a deeper conversation.

A good example to refer to is how metascience should guide us in managing the contribution of AI to science. Lisa Messeri and Molly Crockett summarize this contribution in four types of tools (2024): (1) "oracles," or tools that can guide and orient us in summarizing a debate and formulating new questions; (2) "surrogates," or tools that allow us to simulate scenarios we could not test in the real world because they are practically or theoretically too complex: an example is (broadly) digital twins, which we discussed in Chapter 1; (3) "quants," or tools with intelligence different from and more powerful (in certain respects) than human abilities, which allow us to make a leap in the type of analysis we can produce: an example we saw in Chapter 4 is the Lean project; (4) "arbiters" that allow us, for example, to evaluate articles or select research projects, as conjectured in Chapter 3. Messeri and Crockett highlight the risk that the unreflective use of such AI tools could create a "monoculture of knowledge" where all problems are solved according to the same logic, dictated by the tool rather than the research context.

The analysis presented in this volume suggests an antidote to the problem raised by Messeri and Crockett. We have built this antidote gradually, piecing together many different parts of a mosaic of contemporary science. That mosaic indicates that it is not only fruitful but also crucial—from a metascientific perspective—to consider how the encounter between scientific problems and technologies occurs (especially in the case of deep tech) and to reflect on how this encounter can and should be managed through scalability. As people who intend to contribute to science, we must not stop at the observation of the existence of certain technologies as tools for research: we must ask ourselves how to invest in developing future tools that innovate research sustainably—tools associated with narratives that can be clearly and explicitly communicated to those who must use them and those who must judge their use; tools that wisely guide the complex relationship between contextual and theoretical-scientific knowledge; tools that manage risk and the last mile by learning from past experiences and thinking about the context of use. Only by adopting this perspective can we hope to avoid the "monoculture of knowledge" from which Messeri and Crockett warn us.

Finally, for individual scientists, the landscape of new digital tools presents an unprecedented opportunity for innovation and efficiency, but also a responsibility to engage thoughtfully. We encourage researchers not to view these technologies from afar, but to push their boundaries, and to adapt them to the unique demands of their specific research contexts. This active engagement means exploring how machines can augment human creativity, accelerate data analysis, generate and test novel hypotheses, and hopefully challenge existing paradigms. It also means participating in the development of new AI-driven tools and sharing insights on best practices and potential pitfalls. By becoming active participants in the evolution of *AI for science*, researchers can help ensure these tools remain servants to discovery, not masters of it.

But beyond adapting individual tools, what broader, systemic actions can scientists take to foster a new research ecosystem? The framework outlined by David Donoho (2024), which we discussed in Chapter 4, seems promising. Devising strategies to bring the scientific community to adopt radical transparency and reproducibility seems most needed. To achieve this goal in *each and every* scientific field, *on a global scale*, may be utopistic. Yet, again, it may be important to start from some fields where the result seems most urgent and most readily obtainable.

The vision of Donoho has indeed gained traction with the rise of platforms that frame discovery as a series of well-defined, *scorable tasks*. For example, Kaggle is an online platform that hosts machine learning competitions. Participants are given a dataset and a well-defined task—such as predicting house prices or classifying medical images—and are ranked on a public leaderboard based on the accuracy of their submissions. The platform emphasizes performance over interpretation, rewarding optimization rather than understanding. It has become a standard environment for benchmarking models, sharing code, and, increasingly, shaping how problems are framed. Similar platforms have emerged in scientific domains. Dream Challenges, DrivenData, AIcrowd, and others apply the competition format to tasks such as disease prediction, molecule generation, and environmental modeling. A dataset is published, a metric is selected, and the scientific process is reformulated as a solvable, scorable task.

This approach can certainly accelerate technical progress and draw in diverse contributors, yet it risks reframing science itself by shifting the emphasis from open-ended inquiry to measurable outcomes. The danger is that science becomes defined not by its foundational questions, but by the architecture of the competition. Of course, as with all technology, the tools we build must keep us tethered to the problems at hand and deepen, not diminish, our understanding. They must not devolve into mere distractions or devices that promise assistance only to command our focus.

In conclusion, governments and institutions, society, and scientists are three macro-categories that outline different perspectives for practically implementing how to make the intertwining of science and technology scalable. Each of these macro-categories could naturally be subdivided, identifying more specific needs and reasons, as well as appropriate strategies to try to balance the different voices and opinions. How do we bring them together? Who is responsible for defining the rules of metascience, and according to what criteria should they be established? Adequately addressing these questions requires a framework for *deliberation* that is both effective and fair—one designed for a global society and capable of sustaining public trust in research. Of course, to develop such a framework with the rigor it deserves is a task that cannot be improvised.

Here, at the Columns of Hercules that mark the boundaries of our inquiry, we conclude this book. We leave these discussions as both a challenge and an invitation to the community of metascience entrepreneurs.

and according to what rationale should they be established? Adequately addressing these questions requires a framework for agreements that is both effective and legitimately designed for a global society and capable of sustaining public trust in research. Of course, to develop such a framework with the rigor it deserves is a task that cannot be improvised.

Here, at the Columns of Hercules that mark the boundaries of our inquiry, we conclude this book. We leave these questions as both a challenge and an invitation to the community of metascience entrepreneurs.

References

Akbashev, A. R., & Kalinin, S. V. (2023). Tackling overpublishing by moving to open-ended papers. *Nature Materials*, *22*(3), 1–2.

Andersen, M. Z., Fonnes, S., & Rosenberg, J. (2021). Time from submission to publication varied widely for biomedical journals: A systematic review. *Current Medical Research and Opinion*, *37*(6), 985–993.

Anstine, D. M., & Isayev, O. (2023). Generative models as an emerging paradigm in the chemical sciences. *Journal of the American Chemical Society*, *145*(16), 8736–8750.

Arora, S., & Barak, B. (2009). *Computational complexity: A modern approach*. Cambridge University Press.

Aspen Institute. (2022). *In favor of pure science: A global report by the Aspen Institute Science & Society Program in collaboration among the 13 international partners and 1 international initiative of the Aspen Institute*. The Aspen Institute.

Atherton, L. (2023, April 20). *What is DARPA?* Popular Science. https://www.popsci.com/technology/what-is-darpa/

Baffy, G., Burns, M. M., Hoffmann, B., Ramani, S., Sabharwal, S., Borus, J. F., Pories, S., Quan, S. F., & Ingelfinger, J. R. (2020). Scientific authors in a changing world of scholarly communication: What does the future hold? *The American Journal of Medicine*, *133*(1), 26–31.

Baldwin, M. (2015). *Making "nature": The history of a scientific journal*. University of Chicago Press.

Beaulieu, L. (1999). Bourbaki's art of memory. *Osiris*, *14*, 219–251.

Berners-Lee, T., & Fischetti, M. (1999). *Weaving the web*. Harper.

Bianconi, G. (2018). *Multilayer networks: Structure and function*. Oxford University Press.

Borges, J. L. (1975). *A universal history of infamy* (N. T. de Giovanni, Trans.). Penguin Books. (Original work published 1935).

Borghini, A. (2015). What is a recipe? *Journal of Agricultural and Environmental Ethics*, *28*(4), 719–738.

Bran, A. M., Adams, C., Li, S. D., Tang, H., Nandy, A., Li, W., Sun, R., Yu, C., Li, R., Wu, R., Sun, S., Chen, M., Yang, S., Lin, M., Zhang, H., Cui, Y., Zare, R. N., & Wei, J. (2023). *ChemCrow: Augmenting large-language models with chemistry tools*. arXiv:2304.05376 [physics.chem-ph].

Calvin, M. (1962). *The impurity of science* (Robbins Lecture, Pomona). U.S. Department of Commerce.

Carleo, G., Cirac, I., Cranmer, K., Daudet, L., Schuld, M., Tishby, N., Vogt-Maranto, L., & Zdeborová, L. (2019). Machine learning and the physical sciences. *Review of Modern Physics, 91*(4), 045002.

Casati, R. (2022). *Philosophy as conceptual negotiation*. HAL. ffhal-03937758f.

Choudhary, A., Fox, G., & Hey, T. (2023). *Artificial intelligence for science: A deep learning revolution*. World Scientific.

Collard, J., de Paiva, V., Fong, B., & Subrahmanian, E. (2022). *Extracting mathematical concepts from text*. arXiv:2208.13830 [cs.CL].

Collodi, C. (2021). *The adventures of Pinocchio* (J. Hooper & A. Kraczyna, Trans.). Penguin. (Original work published 1883).

COVIDSurg Collaborative, & GlobalSurg Collaborative. (2021). SARS-CoV-2 vaccination modelling for safe surgery to save lives: Data from an international prospective cohort study. *British Journal of Surgery, 108*(9), 1056–1063.

Cubitt, T., Perez-Garcia, D., & Wolf, M. (2015). Undecidability of the spectral gap. *Nature, 528*, 207–211.

Damasio, A. R. (1994). *Descartes' error: Emotion, reason, and the human brain*. Putnam.

Denton, P. B., Parke, S. J., Tao, T., & Zhang, X. (2022). Eigenvectors from eigenvalues: A survey of a basic identity in linear algebra. *Bulletin of the American Mathematical Society, 59*(1), 31–58.

Diggins, F. W. (1999). The true history of the discovery of penicillin, with refutation of the misinformation in the literature. *British Journal of Biomedical Science, 56*(2), 83–93.

Donoho, D. (2024). Data science at the singularity. *Harvard Data Science Review, 6*(1). https://doi.org/10.1162/99608f92.b91339ef

Dunne, E. (2019). Looking at the mathematics literature. *Notices of the American Mathematical Society, 66*(2), 227–230.

Eco, U. (2004, July). Credere alla tradizione. *L'Espresso*.

Einstein, A. (1934). On the method of theoretical physics. *Philosophy of Science, 1*(2), 163–169.

Einstein, A. (1935–1936, Winter). Science and society. *Science*. Reprinted in A. Einstein, *Essays in humanism*, Open Road Media, 2011.

Ericsson, K. A., Krampe, R. T., & Tesch-Römer, C. (1993). The role of deliberate practice in the acquisition of expert performance. *Psychological Review, 100*(3), 363–406.

Fleming, N. (2021). Sidelined: How to tackle authorship disputes. *Nature, 594*(7862), S1–S2.

Fortunato, S., Bergstrom, C. T., Börner, K., Evans, J. A., Helbing, D., Milojević, S., Petersen, A. M., et al. (2018). Science of science. *Science, 359*(6379), eaao0185.

Foucault, M. (1988). *Technologies of the self*. University of Massachusetts Press.

Freedman, D. H. (2010). *Wrong: Why experts keep failing us—And how to know when not to trust them*. Little, Brown and Company.

Future House. (2023). Retrieved May 20, 2024, from https://www.futurehouse.org

GitHub. (2022). *leanprover-community*. Retrieved May 20, 2024, from https://github.com/leanprover-community

Gladwell, M. (2008). *Outliers: The story of success*. Little, Brown and Company.

Gough, M. (Ed.). (2003). *Politicizing science: The alchemy of policymaking*. Hoover Institution Press.

Gourévitch, A., Portincaso, M., de la Tour, A., Goeldel, N., & Chaudhry, U. (2021, March 11). *Deep tech and the great wave of innovation*. Boston Consulting Group. https://www.bcg.com/publications/2021/deep-tech-innovation

Grover, L. K. (1996). A fast quantum mechanical algorithm for database search. *Proceedings of the twenty-eighth annual ACM symposium on Theory of Computing*, 212–219.

Hagve, M. (2020, August 17). The money behind academic publishing. *Tidsskrift for Den norske legeforening, 140*(11). https://doi.org/10.4045/tidsskr.20.0118

Hardy, G. H. (1992). *A mathematician's apology*. Cambridge University Press. (Original work published 1941).

Harzing, A.-W. (2007). *Publish or perish*. https://harzing.com/resources/publish-or-perish

Haufe, C. (2022). *How knowledge grows: The evolutionary development of scientific practice*. MIT Press.

Ilievski, F., O'Connor, A., Garijo, D., & Szekely, P. (2020). KGTK: A toolkit for large knowledge graph manipulation and analysis. In *International Semantic Web Conference* (pp. 278–293). Springer.

Immerman, N. (1999). *Descriptive complexity*. Springer-Verlag.

Ioannidis, J. P. A. (2018, October). Science funding is broken. The way we pay for science does not encourage the best results. *Scientific American*. https://www.scientificamerican.com/article/science-funding-is-broken/

JWST. (2022). *James Webb space telescope*. Retrieved December 11, 2025, from https://jwst.nasa.gov/index.html

Kitano, H. (2021). Nobel Turing Challenge: Creating the engine for scientific discovery. *npj Systems Biology and Applications*, *7*(1), 29.

Kitcher, P. (2001). *Science, truth, and democracy*. Oxford University Press.

Kline, S. J. (1985). What is technology? *Bulletin of Science, Technology & Society*, *5*(3), 215–218.

Knuth, D. E. (1974). Computer programming as an art. *Communications of the ACM, 17*(12), 667–673.

Koch, K. (2013, February 12). *New ways to fund science*. The Harvard Gazette. https://news.harvard.edu/gazette/story/2013/02/new-ways-to-fund-science/

Kramer, S., Cerrato, M., Džeroski, S., & King, R. (2023). *Automated scientific discovery: From equation discovery to autonomous discovery systems*. arXiv:2305.02251 [cs.AI].

Krenn, M., Aiello, F., Aspuru-Guzik, A., Bäck, T., Bogojeska, J., Bruna, J., Carleo, G., Ceriotti, M., Cussat-Blanc, S., Daudet, L., Delvenne, J. C., Dunjko, V., Eckmann, J. P., Escher, S., Fu, W., Gatti, M., Gurevich, D., Helbling, F., Hutter, F., & Zdeborová, L. (2022). On scientific understanding with artificial intelligence. *Nature Reviews Physics*, *4*(11), 761–769.

Krumholz, H. M. (2015). The end of journals. *Circulation: Cardiovascular Quality and Outcomes*, *8*(6), 533–534.

Lee, C. J., Sugimoto, C. R., Zhang, G., & Cronin, B. (2013). Bias in peer review. *Journal of the American Society for information Science and Technology*, *64*(1), 2–17. https://doi.org/10.1002/asi.22784

Li, M., & Vitányi, P. (2013). *An introduction to Kolmogorov complexity and its applications* (2nd ed.). Springer-Verlag.

Li-Hua, R. (2007). What is technology? *Journal of Technology Management in China, 2*(3), 223–233.

Llorens, A., Tzovara, A., Bellier, L., Bhaya-Grossman, I., Bidet-Caulet, A., Chang, W.K., Cross, Z. R., et al. (2021). Gender bias in academia: A lifetime problem that needs solutions. *Neuron, 109*(13), 2047–2074.

Lunden, I. (2020, March 11). *What do we mean when we talk about deep tech?* TechCrunch. https://techcrunch.com/2020/03/11/what-do-we-mean-when-we-talk-about-deep-tech/

Megill, N., & Wheeler, D. (2019). *Metamath: A computer language for mathematical proofs*. Lulu Press.

Mehlenbacher, A. R. (2022). *On expertise: Cultivating character, goodwill, and practical wisdom*. Penn State University Press.

Messeri, L., & Crockett, M. J. (2024). Artificial intelligence and illusions of understanding in scientific research. *Nature, 627*, 49–58.

Millikan, R. A. (1924). The practical value of pure science. *Science, 59*(1514), 7–10.

Moore, J. P. (2007). Speaking out about US science output. *Science, 318*(5852), 913–913.

Morris, E. M. (2014). *What is 'technology'?* (Indiana University Robert H. McKinney School of Law Research Paper No. 2014–1). https://papers.ssrn.com/sol3/papers.cfm?abstract_id=2377319

Mozzochi, C. (2006). *The Fermat proof*. Trafford Publishing.

Munari, B. (1937, July). Che cosa sono le macchine inutili e perché. *La Lettura, 37*(7), 660–665.

Nielsen, M. (2011). *Reinventing discovery: The new era of networked science*. Princeton University Press.

Nielsen, M. (2025). ASI existential risk: Reconsidering alignment as a goal. April 14. Retrieved August 27, 2025, from https://michaelnotebook.com/xriskbrief/index.html

Nielsen, M., & Qiu, K. (2022). *A vision of metascience*. Retrieved May 20, 2024, from https://scienceplusplus.org/metascience

OECD. (2015). *Frascati manual 2015: Guidelines for collecting and reporting data on research and experimental development*. https://doi.org/10.1787/9789264239012-en

Ong, S. P. (2023, November). *Why Nature Portfolio should have never published the Google AI for materials* [LinkedIn post]. LinkedIn.

Perelman, G. (2003). *Finite extinction time for the solutions to the Ricci flow on certain three-manifolds*. arXiv:math.DG/0307245.

Pievani, T. (2024). *Serendipity: The unexpected in science*. MIT Press.

Piper, K. (2021, December 18). *Can a new approach to funding scientific research unlock innovation?* Vox. https://www.vox.com/future-perfect/2021/12/18/22838746/biomedicine-science-grants-arc-institute

PitchBook & National Venture Capital Association. (2024). *Q4 2023 PitchBook-NVCA Venture Monitor*. https://nvca.org/sdm_downloads/q4-2023-pitchbook-nvca-venture-monitor/

Poincaré, H. (1908). L'avenir des mathématiques. *Revue Générale des Sciences Pures et Appliquées, 19*(930), 39–41.

Politzer-Ahles, S., Girolamo, T., & Ghali, S. (2020). Preliminary evidence of linguistic bias in academic reviewing. *Journal of English for Academic Purposes, 47*, 100895.

Raatikainen, P. (2022). Gödel's incompleteness theorems. In E. N. Zalta & U. Nodelman (Eds.), *The Stanford encyclopedia of philosophy* (Winter 2022 ed.). Stanford University. https://plato.stanford.edu/archives/win2022/entries/goedel-incompleteness/

Rossi, P. (1962): *I filosofi e le macchine 1400 – 1700*. Feltrinelli.

Sandel, M. J. (2020). *The tyranny of merit: What's become of the common good?* Farrar, Straus and Giroux.

Sarewitz, D. (2016). Saving science. *The New Atlantis, 49*, 4–40.

Scientific American. (2011, May 1). *Dr. No Money: The broken science funding system*. https://www.scientificamerican.com/article/dr-no-money/

Shapiro, S. (2000). *Thinking about mathematics*. Oxford University Press.

Stuss, D. T., & Benson, D. F. (1986). *The frontal lobes*. Raven Press EBooks.

Surya, S. (2019). The causal set approach to quantum gravity. *Living Reviews in Relativity, 22*(1), 5.

Tao, T. (2007). What is good mathematics? *Bulletin of the American Mathematical Society, 44*(4), 623–634.

Teller, E., & Brown, A. (1962). *The legacy of Hiroshima*. Doubleday.

Tiberio, V. (1895). Sugli estratti di alcune muffe. *Annali d'Igiene Sperimentale, 5*, 91–103.

Tymoczko, T. (1979). The four-color problem and its philosophical significance. *Journal of Philosophy, 76*(2), 57–83.

Vogels, E., Rainie L., & Anderson J. (2020, June 30). *Experts predict more digital innovation by 2030 aimed at enhancing democracy*. Pew Research Center. https://www.pewresearch.org/internet/2020/06/30/experts-predict-more-digital-innovation-by-2030-aimed-at-enhancing-democracy/

Wang, D., & Barabási, A.-L. (2021). *The science of science*. Cambridge University Press.

Weisskopf, V. F. (1988). *The privilege of being a physicist*. W. H. Freeman & Co.

Wiedijk, F. (2007). *The QED Manifesto revisited*. In R. Matuszwski & A. Zalewska (Eds.), *From insight to proof* (pp. 121–133). Festschrift in Honour of Andrzej Trybulec.

Wilczek, F. (2006). *Fantastic realities: 49 Mind journeys and a trip to Stockholm*. World Scientific Publishing.

Wolchover, N. (2019, November 13). Neutrinos lead to unexpected discovery in basic math. *Quanta Magazine*.

Wooldridge, M. (2009). *An introduction to multiagent systems* (2nd ed.). John Wiley & Sons.

Youngman, R. (2022). *The contribution of physics to the UK economy*. Institute of Physics. Retrieved May 20, 2024, from https://www.iop.org/strategy/productivity-programme/physics-and-economy

Zuboff, S. (2019). *The age of surveillance capitalism: The fight for a human future at the new frontier of power*. Profile Books.

Zuckerman, G. (2023, October 4). After shunning scientist, University of Pennsylvania celebrates her Nobel Prize. *The Wall Street Journal*. https://web.archive.org/web/20241007010105/https://www.wsj.com/health/after-shunning-scientist-university-of-pennsylvania-celebrates-her-nobel-prize-96157321

Index

For Product Safety Concerns and Information please contact our EU representative GPSR@taylorandfrancis.com Taylor & Francis Verlag GmbH, Kaufingerstraße 24, 80331 München, Germany